G. Prodi (Ed.)

Eigenvalues of Non-Linear Problems

Lectures given at a Summer School of the
Centro Internazionale Matematico Estivo (C.I.M.E.),
held in Varenna (Como), Italy,
June 16-25, 1974

C.I.M.E. Foundation
c/o Dipartimento di Matematica "U. Dini"
Viale margagni n. 67/a
50134 Firenze
Italy
cime@math.unifi.it

ISBN 978-3-642-10939-3 e-ISBN: 978-3-642-10940-9
DOI:10.1007/978-3-642-10940-9
Springer Heidelberg Dordrecht London New York

Reprint of the 1st ed. C.I.M.E., Ed. Cremonese, Roma 1974
With kind permission of C.I.M.E.

Printed on acid-free paper

Springer.com

CENTRO INTERNAZIONALE MATEMATICO ESTIVO

(C. I. M. E.)

III Ciclo - Varenna dal 16 al 25 giugno 1974

EIGENVALUES OF NON-LINEAR PROBLEMS

Coordinatore: Prof. G. PRODI

H. AMANN	: Nonlinear eigenvalue problems in ordered Banach spaces.	pag.	1
P. C. FIFE	: Branching phenomena in fluid dynamics and chemical reaction-diffusion theory.	"	23
W. KLINGENBERG	: The theory of closed geodesics.	"	85
P. H. RABINOWITZ	: Variational methods for nonlinear eigenvalue problems.	"	139
M. REEKEN	: Existence of solutions to the hartree-fock equations.	"	197
R. E. L. TURNER	: Positive solutions of nonlinear eigenvalue problems.	"	211

CENTRO INTERNAZIONALE MATEMATICO ESTIVO

(C. I. M. E.)

NONLINEAR EIGENVALUE PROBLEMS IN ORDERED BANACH SPACES

H. AMANN

Corso tenuto a Verenna dal 16 al 25 giugno 1974

NONLINEAR EIGENVALUE PROBLEMS
IN ORDERED BANACH SPACES

Herbert Amann

1. Introduction

In this paper we study nonlinear elliptic boundary value problems of the form

$$
(1.1) \quad \begin{aligned} Lx(t) &= \lambda\phi(t,x(t)) && \text{in } \Omega , \\ Bx(t) &= o && \text{on } \partial\Omega , \end{aligned}
$$

where Ω is a bounded smooth domain in $\mathbb{R}^N$, $N \geq 1$, L is a second order strongly uniformly elliptic real differential operator, B is an at most first order real boundary operator, and λ is a real number.

As for the pair (L,B) , we impose the following fundamental hypotheses:

(H1) *The pair* (L,B) *satisfies the strong maximum principle,* that is, for every function $x \in C^2(\Omega) \cap C^1(\overline{\Omega})$ such that

$$
\begin{aligned} Lx(t) &\geq o && \text{in } \Omega , \\ Bx(t) &\geq o && \text{on } \partial\Omega , \end{aligned}
$$

it follows that $x(t) \geq o$ on $\overline{\Omega}$. Moreover, if $x \not\equiv o$ then $x(t) > o$ for every $t \in \Omega$, and for every $t \in \partial\Omega$ such that $x(t) = o$, it follows that $\frac{\partial x}{\partial \nu}(t) < o$, where ν denotes the outer normal on $\partial\Omega$.

(H2) *The pair* (L,B) *satisfies Schauder type a priori estimates,* that is, there exists a constant $\gamma > o$ such that for every $x \in C^{2+\mu}(\overline{\Omega})$ satisfying the boundary conditions $Bx(t) = o$ on $\partial\Omega$,

$$\|x\|_{C^{2+\mu}(\overline{\Omega})} \leq \gamma \|Lx\|_{C^{\mu}(\overline{\Omega})},$$

where $\mu \in (o,1)$ if $N \geq 2$, and $\mu = o$ if $N = 1$.

It is well known that hypothesis (H2) is satisfied if the coefficients of L belong to $C^{\mu}(\overline{\Omega})$, $Bx = x|\partial\Omega$, and the homogeneous problem

$$Lx = o \; , \; Bx = o$$

has the trivial solution only. Clearly, this latter condition is satisfied if hypothesis (H1) holds. If B is the Dirichlet boundary operator, that is, $Bx = x|\partial\Omega$, then (H1) is satisfied if the coefficient of the non-differentiated term in L is nonnegative. For more general boundary conditions we refer to [1,2].

As for the nonlinearity ϕ, we impose the following hypothesis.

(H3) (i) $\phi \in C^2(\overline{\Omega} \times \mathbb{R}_+)$,

(ii) *there exist positive constants* γ_0 *and* δ *such that for every* $(t,\xi) \in \overline{\Omega} \times \mathbb{R}_+$, $\phi(t,\xi) \geq \gamma_0 + \delta\xi$,

(iii) *there exists a nonnegative constant* ω_0 *such that for every* $(t,\xi) \in \overline{\Omega} \times \mathbb{R}_+$, $D_2\phi(t,\xi) > -\omega_0$, where D_2 denotes the partial derivative with respect to the second variable.

It should be remarked that for simplicity we restrict our considerations to nonlinear eigenvalue problems of the form (1.1). However, it is easily seen that the same methods apply to more general problems of the form

$$\begin{aligned} Lx(t) &= \psi(t,x(t),\lambda) \quad \text{in} \quad \Omega , \\ Bx(t) &= \chi(t) \quad \text{on} \quad \partial\Omega . \end{aligned}$$

It is an immediate consequence of the above hypotheses that for the solvability of (1.1) it is necessary that $\lambda \geq o$. Moreover, if for some $\lambda > o$, problem (1.1) has a solution x then $x \neq o$ and $x(t) \geq o$ for $t \in \overline{\Omega}$. In order to use this important information we transform the nonlinear elliptic eigenvalue problem (1.1) into an abstract fixed point equation depending on a real parameter in a suitable ordered Banach space.

Hypothesis (H2) implies that L has a continuous inverse L^{-1} mapping $C^{\mu}(\overline{\Omega})$ into $C^{2+\mu}(\overline{\Omega})$. It follows from (H1) that L^{-1} is a *positive linear operator*, that is, the images of nonnegative functions are nonnegative functions. Hence it seems natural to transform problem

(1.1) into a fixed point equation in the Banach space $C^{\mu}(\overline{\Omega})$ where this space is given the natural ordering. However, for technical reasons (the positive cone is not normal) we do not use the space $C^{\mu}(\overline{\Omega})$ but $C(\overline{\Omega})$. In fact, it will be necessary to consider a certain subspace of $C(\overline{\Omega})$ which is particularly well adapted to the differential equation.

Let T be a nonempty set and let $x : T \to \mathbb{R}$ be a function on T . Then we write $x \geq o$ if $x(t) \geq o$ for every $t \in T$, and $x > o$ if $x \geq o$ but $x \neq o$. In the latter case, x is said to be *positive.*

It is well known that for every $y \in C^{\mu}(\overline{\Omega})$ the linear boundary value problem (BVP)

$$Lx = y \quad \text{in} \quad \Omega ,$$

$$Bx = o \quad \text{on} \quad \partial\Omega ,$$

has a unique solution $x := Ky$ in $C^{2+\mu}(\overline{\Omega})$. By using the L_p-theory for elliptic BVPs and Sobolev type imbedding theorems it can be shown (e.g. [1]) that the linear operator K defined above has a continuous extension, denoted again by K , to a compact linear operator $K : C(\overline{\Omega}) \to C^{1}(\overline{\Omega})$, the *solution operator (for the pair* (L,B)).

It is a consequence of the strong maximum principle that the solution operator K is not only a positive linear operator but it maps every positive function into a function without zeros in Ω . The following lemma (which is proved in [1]) contains the precise statement of this fact in a form which can be generalized to abstract ordered Banach spaces.

(1.1) Lemma: The solution operator K *maps* $C(\bar{\Omega})$ *compactly into* $C^1(\bar{\Omega})$ *and* K *is* e_0*-positive, that is, there exists a positive function* $e_0 \in C(\bar{\Omega})$ *such that for every* $x \in C(\bar{\Omega})$ *with* $x > o$ *, there are positive numbers* α *and* β *such that*

$$\alpha e_0 \le Kx \le \beta e_0 \quad . \tag{1.2}$$

For e_0 *we may take the unique solution of the linear* BVP

$$Lx = \mathbb{1} \quad \text{in} \quad \Omega ,$$
$$Bx = o \quad \text{on} \quad \partial\Omega ,$$

where $\mathbb{1}(t) = 1$ *for every* $t \in \Omega$.

In particular, Lemma (1.1) implies that K can be considered as a positive compact endomorphism of $C(\bar{\Omega})$ and there exists a positive number α such that $Ke_0 \ge \alpha e_0$. Consequently, the Krein-Rutman theorem [6] applies and it follows that the spectral radius $r(K)$ is positive and an eigenvalue of K having a positive eigenfunction x_0 . It is easily seen that $\lambda_0 := r(K)^{-1}$ is the smallest eigenvalue, the *principal eigenvalue*, of the linear eigenvalue problem

$$\begin{aligned} Lx &= \lambda x \quad \text{in} \quad \Omega , \\ Bx &= o \quad \text{on} \quad \partial\Omega , \end{aligned} \tag{1.3}$$

and x_0 is a positive eigenfunction of (1.3).

We denote by $C_+(\overline{\Omega})$ the *positive cone* in $C(\overline{\Omega})$, that is, $C_+(\overline{\Omega})$ consists of o and all positive continuous functions on $\overline{\Omega}$. Then we define a continuous and bounded map

$$F : C_+(\overline{\Omega}) \to C_+(\overline{\Omega})$$

by

$$F(x)(t) := \phi(t,x(t)) \quad \text{for} \quad x \in C_+(\overline{\Omega}) \text{ and } t \in \overline{\Omega} ,$$

that is, F is the Nemytskii operator for the function ϕ . Then it is easily verified that the nonlinear elliptic eigenvalue problem (1.1) is equivalent to the fixed point equation

$$x = \lambda KF(x)$$

on $C_+(\overline{\Omega})$.

It is an immediate consequence of hypothesis (H3(ii)) that for every $x \in C_+(\overline{\Omega})$,

$$KF(x) \geq \gamma_0 K\mathbb{1} + \delta Kx = \gamma_0 e_0 + \delta Kx .$$

This inequality can be used to prove the following nonexistence theorem.

<u>*(1.2) Theorem:*</u> *Denote by* λ_0 *the principal eigenvalue of the linear eigenvalue problem* (1.3). *Then the nonlinear elliptic* BVP (1.1) *has no solution if* $\lambda \geq \lambda_0/\delta$.

Proof: By the Krein-Rutman theorem the dual operator K' has an eigenvector x^* to the eigenvalue $r(K) = \lambda_0^{-1}$ such that $(x^*,x) \geq o$ for every $x \in C_+(\overline{\Omega})$.

We claim that for every $x \in \dot{C}_+(\bar{\Omega}) := C_+(\bar{\Omega}) \setminus \{o\}$, $(x^*,x) > o$. Indeed, since K is e_0-positive, inequalities (1.2) imply that

$$\alpha(x^*,e_0) \le (x^*,x) \le \beta(x^*,e_0) \quad , \tag{1.4}$$

where α and β are positive numbers depending on $x \in \dot{C}_+(\bar{\Omega})$. Suppose that $(x^*,e_0) = o$. Then the preceding inequalities imply that $(x^*,x) = o$ for every $x \in C_+(\bar{\Omega})$. Since every $x \in C(\bar{\Omega})$ can be represented in the form $x = y - z$ with $y,z \in C_+(\bar{\Omega})$, it follows that $(x^*,x) = o$ for every $x \in C(\bar{\Omega})$. Thus $x^* = o$, which contradicts the fact that x^* is an eigenvector. Consequently, $(x^*,e_0) > o$ and by (1.4), $(x^*,x) > o$ for every $x > o$.

Now suppose that for some $\lambda > o$ the eigenvalue problem (1.1) has a solution x , or, equivalently, that $x = \lambda KF(x)$. Since $F(x)$ is positive and K is e_0-positive, it follows that $x > o$. Moreover,

$$x = \lambda KF(x) \ge \lambda\gamma_0 e_0 + \lambda\delta Kx \quad .$$

By applying to this inequality the functional x^* , we obtain

$$(x^*,x) \ge \lambda\gamma_0(x^*,e_0) + \lambda\delta(x^*,Kx) > \lambda\delta r(K)(x^*,x) \; .$$

Thus $\lambda\delta r(K) < 1$, that is, $\lambda < \lambda_0/\delta$. Q.E.D.

So far we have essentially only used the fact that the solution operator K is positive. It is to be expected that we can obtain better results if the full nonlinear map λKF is compatible with the ordering of the underlying space. By using the a priori bound of Theorem (1.2) we can achieve such a situation by means of the following simple device.

Let $\omega := \lambda_0\omega_0/\delta$. Thenproblem (1.1) is obviously equivalent to the nonlinear BVP

$$(L + \omega)x(t) = \lambda\phi(t,x(t)) + \omega x(t) \quad \text{in} \quad \Omega ,$$
$$Bx(t) = o \qquad\qquad \text{on} \quad \partial\Omega .$$

Denote by K_ω the solution operator for the pair $(L + \omega, B)$ and define

$$F_\omega : \mathbb{R}_+ \times C_+(\bar{\Omega}) \to C_+(\bar{\Omega})$$

by

$$F_\omega(\lambda,x) := \lambda F(x) + \omega x \quad .$$

Then the nonlinear elliptic eigenvalue problem (1.1) is equivalent to the fixed point equation

$$x = K_\omega F_\omega(\lambda,x) .$$

Since $(L + \omega, B)$ satisfies hypotheses (H1) and (H2) it follows again that K_ω is an e-positive compact linear operator from $C(\bar{\Omega})$ into $C^1(\bar{\Omega})$, where $e := K_\omega \mathbf{1}$. Since F_ω is continuous and bounded, the map

$$f : \mathbb{R}_+ \times C_+(\bar{\Omega}) \to C_+(\bar{\Omega})$$

defined by

$$f(\lambda,x) := K_\omega F_\omega(\lambda,x)$$

is completely continuous, that is, f is continuous and maps bounded sets into compact sets.

By Theorem (1.2) it suffices to consider the restriction of f to the set $[o,\lambda_0/\delta) \times C_+(\bar{\Omega})$. On this set the map f has the important property that it is *increasing*, that is, for every pair (λ,x) , $(\mu,y) \in [o,\lambda_0/\delta) \times C_+(\bar{\Omega})$ such that $\mu \le \lambda$ and $y \le x$,

$$f(\mu,y) \le f(\lambda,x) \quad . \tag{1.5}$$

Indeed, for every such pair (λ,x) , (μ,y) , the positivity of K_ω implies the inequality

$$
\begin{aligned}
& f(\lambda,x) - f(\mu,y) = \\
& (\lambda - \mu)K_\omega F(x) + \mu K_\omega(F(x) - F(y) + \mu^{-1}\omega(x - y)) \geq \\
& (\lambda - \mu)K_\omega F(x) + \mu K_\omega(F(x) - F(y) + \omega_0(x - y)) .
\end{aligned} \tag{1.6}
$$

Hence (1.5) is a consequence of hypothesis (H3).

Now observe that, due to the e-positivity of K_ω , every solution x of the fixed point equation $x = f(\lambda,x)$ is comparable to the function e in the sense that it belongs to the subset

$$C_e(\bar{\Omega}) := \{x \in C(\bar{\Omega}) \mid \text{there exists a positive number } \alpha = \alpha(x) \text{ such that } -\alpha e \leq x \leq \alpha e\}$$

of $C(\bar{\Omega})$. More precisely, f maps all of $[o,\lambda_0/\delta) \times C_+(\bar{\Omega})$ into

$$C_{e,+}(\bar{\Omega}) := C_e(\bar{\Omega}) \cap C_+(\bar{\Omega}) .$$

Hence it suffices to consider the restriction f_e of f to $[o,\lambda_0/\delta) \times C_{e,+}(\bar{\Omega})$.

Clearly, $C_e(\bar{\Omega})$ is a vector subspace of $C(\bar{\Omega})$. It can be made into a Banach space by means of the e-*norm*,

$$\|x\|_e := \inf\{\alpha > o \mid -\alpha e \leq x \leq \alpha e\} .$$

The e-norm is stronger than the maximum norm and it is *monotone*, that is, $\|x\|_e \leq \|y\|_e$ whenever $o \leq x \leq y$. The set $C_{e,+}(\bar{\Omega})$ is the positive cone of $C_e(\bar{\Omega})$, that is, it consists of o and all positive continuous functions belonging to $C_e(\bar{\Omega})$. With respect to the e-norm, $C_{e,+}(\bar{\Omega})$ is closed and has nonempty interior. In fact, x is an interior point of

$C_{e,+}(\bar{\Omega})$ if and only if there exist positive numbers α and β such that $\alpha e \le x \le \beta e$. A proof of these assertions can be found in [5] .

In the following, $C_e(\bar{\Omega})$ will always be given the e-norm. Hence $C_e(\bar{\Omega})$ is a Banach space which is continuously imbedded in $C(\bar{\Omega})$. If $e(t) > o$ for every $t \in \bar{\Omega}$, then $C_e(\bar{\Omega})$ is topologically isomorphic to $C(\bar{\Omega})$. In case of Dirichlet boundary conditions we have $e|\partial\Omega = o$ and it can be shown that $C_0^1(\bar{\Omega}) := \{x \in C^1(\bar{\Omega}) \mid x|\partial\Omega = o\}$ is continuously imbedded in $C_e(\bar{\Omega})$. These facts and the compactness of K_ω as a map from $C(\bar{\Omega})$ into $C^1(\bar{\Omega})$ imply that K_ω maps $C(\bar{\Omega})$ compactly into $C_e(\bar{\Omega})$ (comp. [1]) . Moreover, since K_ω is e-positive, it maps $\dot{C}_+(\bar{\Omega})$ into the interior of $C_{e,+}(\bar{\Omega})$, that is, K_ω is *strongly positive.*

Using these facts and the definition of f_e , it is easily seen that f_e has the following property:

(P1) The map

$$f_e : [o,\lambda_0/\delta) \times C_{e,+}(\bar{\Omega}) \to C_{e,+}(\bar{\Omega})$$

is completely continuous. The map $f_e(o,\cdot)$ has exactly one fixed point, namely $x = o$. There exists $\rho > o$ such that for every positive x with $\|x\|_e = \rho$ and every $\sigma \ge 1$, $f_e(o,x) \neq \sigma x$.

Moreover it follows from inequality (1.6) and the e-positivity of K_ω that the map f_e is *strongly increasing,* that is, for every pair of distinct points (λ,x) , $(\mu,y) \in [o,\lambda_0/\delta) \times C_{e,+}(\bar{\Omega})$ satisfying $\mu \le \lambda$ and $y \le x$,

$$f_e(\lambda,x) - f_e(\mu,y) \in \text{int}\, C_{e,+}(\overline{\Omega}) \, .$$

The regularity hypothesis (H3(i)) implies the following further property of f_e, which, in turn, implies that f_e is strongly increasing.

(P2) The map f_e is twice continuously differentiable on $[o,\lambda_o/\delta) \times C_{e,+}(\overline{\Omega})$. For every $(\mu,y) \in (o,\lambda_o/\delta) \times C_{e,+}(\overline{\Omega})$, the derivative

$$f_e'(\lambda,x) = (D_1 f_e(\lambda,x), D_2 f_e(\lambda,x)) : \mathbb{R} \times C_e(\overline{\Omega}) \to C_e(\overline{\Omega})$$

is a strongly positive continuous linear operator.

These lengthy considerations show that we can use the information which is contained in the maximum principle and the existence and regularity theory for linear elliptic BVPs to transform the nonlinear elliptic eigenvalue problem (1.1) into an equivalent fixed point equation of the form

$$x = f_e(\lambda,x)$$

in a Banach space $C_e(\overline{\Omega})$ such that both, the Banach space and the map enjoy the relatively pleasant properties given above. In this form we can apply the theory of ordered Banach spaces and the methods of nonlinear functional analysis to obtain relatively precise information about the solvability of problem (1.1).

2. Nonlinear Eigenvalue Problems in Ordered Banach Spaces

Let E be a real Banach space. A subset $P \subset E$ is called a cone if P is closed, $P + P \subset P$, $\mathbb{R}_+ P \subset P$, and $P \cap (-P) = \{o\}$. Given a cone P in E , we define an ordering in E by setting $x \leq y$ iff $y - x \in P$. Then (E,P) is called an ordered Banach space (OBS) with positive cone P . The elements $x \in \dot{P} := P \setminus \{o\}$ are called positive and we write $x < y$ to mean that $y - x \in \dot{P}$. Throughout this paper we assume that

> (E,P) *is an* OBS *whose positive cone has nonempty interior* $\overset{o}{P}$ *and whose norm is monotone, that is,* $\|x\| \leq \|y\|$ *whenever* $o \leq x \leq y$.

Observe that the OBSs $(\mathbb{R},\mathbb{R}_+)$ and $(\mathbb{R} \times E, \mathbb{R}_+ \times P)$ enjoy the above properties also, where the latter space is endowed with one of the usual norms, e.g. $\|(\lambda,x)\| = |\lambda| + \|x\|$.

We consider an equation of the form

$$x = f(\lambda,x) \quad , \tag{2.1}$$

where the map f satisfies the following assumptions:

> $f : \mathbb{R}_+ \times P \to P$ *and is completely continuous.*
> *The map* $f(o,\cdot) : P \to P$ *has exactly one fixed point, namely* $x = o$.
> *There exists* $\rho > o$ *such that for every* $x \in P$ *with* $\|x\| = \rho$

every $\sigma \geq 1$, $f(o,x) \neq \sigma x$. *There exists* $\overline{\lambda} > o$ *such that* $f(\lambda,\cdot)$ *has no fixed point for* $\lambda \geq \overline{\lambda}$.

The map f *is twice continuously differentiable on* $[o,\overline{\lambda}) \in P$ *and for every* $(\lambda,x) \in (o,\overline{\lambda}) \times P$, *the derivative*

$f'(\lambda,x) = (D_1 f(\lambda,x) , D_2 f(\lambda,x)) : \mathbb{R} \times E \to E$

is a strongly positive linear operator, that is,

$f'(\lambda,x)((\mathbb{R}_+ \times P)^{\cdot}) \subset \overset{o}{P}$.

We denote by Σ the solution set of equation (2.1), that is,

$$\Sigma := \{(\lambda,x) \in \mathbb{R}_+ \times P \mid x = f(\lambda,x)\} .$$

It follows from the results of Dancer [4] that the solution set is locally compact and contains an unbounded component emanating from (o,o) .

In the following we give more detailed information about the structure of Σ . For this purpose we denote by Λ the projection of Σ into $\mathbb{R}_+$, that is,

$$\Lambda = \{\lambda \in \mathbb{R} \mid f(\lambda,\cdot) \text{ has a fixed point}\} .$$

Since the derivative f' is strongly positive on $(o,\overline{\lambda}) \times P$ it follows that f is strongly increasing on this set, that is, for every pair of distinct points (λ,x) , $(\mu,y) \in (o,\overline{\lambda}) \times P$ such that $(\lambda,x) < (\mu,y)$, it follows that $f(\mu,y) - f(\lambda,x) \in \overset{o}{P}$. By using this fact, it is easy to prove the following theorem.

(2.1) Theorem: Λ *is a nontrivial interval containing* o . *For every* $\lambda \in \Lambda$, *there exists a minimal fixed point* $\overline{x}(\lambda)$ *of* $f(\lambda,\cdot)$.

The map $\overline{x}(\cdot) : \Lambda \to P$ *is strongly increasing and left continuous.*

Clearly, $\overline{x}(\lambda)$ is a minimal fixed point iff every fixed point x_λ of $f(\lambda,\cdot)$ satisfies $x_\lambda \geq \overline{x}(\lambda)$. In particular, the minimal fixed point is unique. In the proof of Theorem (2.1) it is shown that $\overline{x}(\lambda)$ can be computed iteratively, namely $\overline{x}(\lambda) = \lim_{k\to\infty} f^k(\lambda,o)$, that is, $\overline{x}(\lambda)$ is the limit of the sequence (x_k) , where $x_o = o$ and $x_{k+1} = f(\lambda,x_k)$.

We set $\lambda^* := \sup \Lambda$. Then $o < \lambda^* \leq \overline{\lambda}$. Next we impose a further hypothesis, namely the existence of an a priori bound.

(H) *There exist* $\mu \in (o,\lambda^*)$ *and* $\rho > \|\overline{x}(\mu)\|$ *such that there is no* $(\lambda,x) \in \Sigma$ *with* $\lambda \geq \mu$ *and* $\|x\| = \rho$.

It is an easy consequence of hypothesis (H) that Λ is closed, that is, $\lambda^* \in \Lambda$. Consequently, $\lambda^* < \overline{\lambda}$.

Now we claim that for every $\lambda \in [\mu,\lambda^*)$, the map $f(\lambda,\cdot)$ has at least two distinct fixed points. By Theorem (2.1) we know already that for every such λ , there exists a minimal fixed point $\overline{x}(\lambda)$. Suppose now that λ is a point of discontinuity of $\overline{x}(\cdot)$. Then it is easy to show that $x_\lambda := \lim_{\sigma\downarrow\lambda} \overline{x}(\sigma)$ exists and is a fixed point of $f(\lambda,\cdot)$ with $x_\lambda > \overline{x}(\lambda)$. Hence it suffices to consider the case where λ is a point of continuity and $\overline{x}(\lambda)$ is an isolated fixed point of $f(\lambda,\cdot)$. In this case we show that the Leray-Schauder fixed point index of $\overline{x}(\lambda)$ is equal to $+1$.

It is an important consequence of the fact that $\overline{x}(\lambda)$ is the minimal fixed point of $f(\lambda,\cdot)$ that

$$r(D_2f(\lambda,\overline{x}(\lambda))) \leq 1 \quad ,$$

where r denotes the spectral radius. Suppose now that $r(D_2f(\lambda,\overline{x}(\lambda))) < 1$. Then it is an easy consequence of the standard Leray-Schauder degree theory that for every sufficiently small $\rho > o$.

$$d(id - f(\lambda,\cdot) , \overline{x}(\lambda) + B_\rho , o) = 1 \quad , \tag{2.2}$$

where B_ρ denotes the open ball in E about o and radius ρ .

It remains to consider the more difficult case where $\overline{x}(\lambda)$ is an isolated fixed point of $f(\lambda,\cdot)$ with

$$r(D_2f(\lambda,\overline{x}(\lambda))) = 1 \quad .$$

In this case, 1 is a simple eigenvalue of $D_2f(\lambda,\overline{x}(\lambda))$. By means of this knowledge and several applications of the implicit function theorem it is possible to show that in a neighborhood of the point $(\lambda,\overline{x}(\lambda))$, the solution set Σ consists of a smooth curve $(\lambda(\sigma),x(\sigma))$ where $\sigma \in (-\varepsilon,\varepsilon)$ and the map $\sigma \to x(\sigma)$ is strongly increasing. Using this local representation of Σ and the homotopy invariance of the Leray-Schauder degree, it can finally be shown that also in this case relation (2.2) is true.

Now we show that for a sufficiently large open subset U of $\overset{\circ}{P}$ which contains the minimal fixed point of $f(\lambda,\cdot)$, the Leray-Schauder degree vanishes. Then it follows from (2.2) and the additivity property of the Leray-Schauder degree, that there must be at least one more fixed point in U .

Since $\overline{x}(\cdot)$ is strongly increasing, there exists $x_0 \in \overset{\circ}{P}$ such that $\overline{x}(\mu) - x_0 \in \overset{\circ}{P}$. Hence hypothesis (H) and Theorem (2.1) imply that for every $\lambda \geq \mu$, no fixed point of $f(\lambda,\cdot)$ is contained on the boundary of the bounded open subset

$$U := (x_0 + \overset{\circ}{P}) \cap B_\rho$$

of E . Consequently, the Leray-Schauder degree $d(id - f(\lambda,\cdot),U,o)$ is well defined and, due to the homotopy invariance,

$$d(id - f(\lambda,\cdot),U,o) = d(id - f(\overline{\lambda},\cdot),U,o)$$

for every $\lambda \geq \mu$. But since $f(\overline{\lambda},\cdot)$ has no fixed points at all, it follows that

$$d(id - f(\lambda,\cdot),U,o) = o$$

for every $\lambda \geq \mu$.

By this way we obtain the following theorem whose detailed proof is given in [2] .

(2.2) Theorem: Let hypothesis (H) *be satisfied. Then there exists* $\lambda^* > o$ *such that problem* (2.1) *has at least one solution for every* $\lambda \in [o,\lambda^*]$ *, no solution for* $\lambda > \lambda^*$ *, and at least two distinct solutions for* $\mu \leq \lambda < \lambda^*$.

3. Applications to Elliptic Boundary Value Problems

As has been shown in the introduction, the results of the preceding paragraph apply to the elliptic eigenvalue problem (1.1) provided the additional hypothesis (H) can be verified. This means that a priori bounds for the solutions of problem (1.1) have to be established.

In the following we exhibit two cases for which the necessary a priori bounds can be established. Namely, the case of two point boundary value problems of ordinary differential equations and the case of asymptotically linear nonlinearities.

In the remainder of this section we suppose that hypotheses (H1) - (H3) are satisfied.

(3.1) Theorem: Suppose that $N = 1$ *, that is,* L *is a second order ordinary differential operator. Moreover suppose that*

$$\lim_{\xi \to \infty} \frac{\phi(t,\xi)}{\xi} = \infty$$

uniformly in $t \in \overline{\Omega}$ *. Then there exists a positive number* λ^* *such that problem* (1.1) *has no solution for* $\lambda > \lambda^*$ *, at least one solution for* $\lambda = \lambda^*$ *, and at least two positive solutions for* $0 < \lambda < \lambda^*$ *.*

A detailed proof of this theorem is given in [3].

The next theorem applies to elliptic BVPs in any dimensions. But the nonlinearity is supposed to be asymptotically linear.

<u>(3.2) Theorem:</u> *Suppose that there exists* $\phi_\infty \in C^\mu(\bar{\Omega})$ *such that*

$$\phi_\infty(t) = \lim_{\xi \to \infty} D_2\phi(t,\xi)$$

uniformly in $t \in \bar{\Omega}$. *In addition suppose that there exists* $y \in \dot{C}_+(\bar{\Omega})$ *and a constant* $\rho > 0$ *such that for all* $t \in \Omega$ *and all* $\xi \geq \rho$,

$$\phi(t,\xi) - D_2\phi(t,\xi) \cdot \xi \leq -y(t) .$$

Denote by λ_∞ *the principal eigenvalue of the linear eigenvalue problem*

$$Lx = \lambda\phi_\infty x \quad \text{in } \Omega ,$$

$$BBx = 0 \quad \text{on } \partial\Omega .$$

Then there exists a positive number λ^* *such that problem* (1.1) *has no solution for* $\lambda > \lambda^*$ *and at least one solution for* $0 \leq \lambda \leq \lambda^*$. *Furthermore,* $0 < \lambda_\infty < \lambda^*$ *and for every* $\lambda \in (\lambda_\infty, \lambda^*)$, *problem* (1.1) *has at least two distinct solutions.*

For a proof of this theorem as well as for further details and bibliographic references we refer to [2].

H. Amann

Bibliography

[1] H. AMANN : On the number of solutions of nonlinear equations in ordered Banach spaces.
J. Functional Anal. 11 (1972), 346-384.

[2] ————: Multiple positive fixed points of asymptotically linear maps.
J. Functional Anal., to appear

[3] ————: On the number of solutions of asyptotically superlinear two point boundary value problems.
Arch. Rat. Mech. Anal., to appear.

[4] E.N. DANCER: Global solution branches for positive mappings.
Arch. Rat. Mech. Anal. 52 (1973), 181-192.

[5] M.A. KRASNOSEL'SKII: Positive Solutions of Operator Equations.
Groningen: Noordhoff 1964.

[6] M.G. KREIN and M.A. RUTMAN: Linear operators leaving invariant a cone in a Banach space.
Amer.Math. Soc. Translations, Ser.1,10,1-128 (1962).

Mathematisches Institut der
Ruhr-Universität
D 463o Bochum, Germany

CENTRO INTERNAZIONALE MATEMATICO ESTIVO

(C. I. M. E.)

BRANCHING PHENOMENA IN FLUID DYNAMICS AND CHEMICAL REACTION-DIFFUSION THEORY

P. C. FIFE

Corso tenuto a Varenna dal 16 al 25 giugno 1974

BRANCHING PHENOMENA IN FLUID DYNAMICS AND CHEMICAL REACTION-DIFFUSION THEORY*

Paul C. Fife

University of Arizona

INTRODUCTION

Branching phenomena in mathematical physics and biology can best be described as fascinating, prevalent, and difficult. The first description expresses my own viewpoint, which I hope you will share with me and a growing number of mathematicians. As for their prevalence, branching phenomena interpreted in a broad sense, includes a multitude of well-known facets of physical and biological systems, from cell division to the transition to turbulence. In these lectures we stay primarily within the context of viscous fluid flow and chemically reacting and diffusing systems. Finally, as to the difficulty mentioned, it refers both to the construction of adequate mathematical models, and to their analysis. Here we deal only with well accepted models. There have been some profound and elegant results achieved in the branching analysis of these models; and yet in another sense the analysis is as yet only skin-deep, considering the number of questions whose answers are still unknown.

Let me delineate the topics to be covered here. First of all, I shall be talking almost entirely about the branching theory of certain nonlinear partial differential equations; specifically, the Navier-Stokes equations and systems of reaction-diffusion equations, including the

---

*Lectures given at C.I.M.E. Advanced Study Institute "Eigensolutions of Nonlinear Equations", Varenna, June, 1974.

corresponding time-independent cases. In effect, this is not much of a limitation, as there is no doubt that an adequate theory of these equations would include an almost overwhelming wealth of qualitative phenomena. Most of the theory has not yet been developed, despite the huge strides that have been made in the study of these equations. So in a sense we are still on the outside looking in.

Secondly, I shall limit myself by and large to rigorous mathematical results, together with some brief commentary and conjectures regarding unproven results, based somewhat on experimental evidence. This is a significant limitation, and excludes some important work of a numerical and experimental nature.

Even within the confines of the equations treated, I must necessarily exclude many interesting developments; I shall indicate further references to some of them in the text of the lectures. However, the selection of subject matter that I have made appears to me to play a central role in the theory. When several solutions of physical problems exist, as happens in the cases examined in these lectures, an immediate further question is, which of them are stable? Due to time limitation I say relatively little about this important question, although some comments and references are given.

Finally, let me say that these lectures constitute a survey of existing results. There is nothing substantial here that has not appeared or will not appear elsewhere, although the point of view and organization is new in many cases. Proofs are often not given in detail, though their main ideas are usually sketched.

Remarks should be made on future research possibilities in this field. The study of the bifurcation and branching behavior of problems in partial differential equations is a relatively new endeavor, and is rapidly gaining momentum. Most of the results described herein are on the frontier of this rapidly expanding subject. Consider, for example, that well over half of the references listed in the bibliography of these lectures were published during the 1970's, and some of the important ones have not yet appeared in print at all. A willing researcher will have no difficulty in finding topics to involve himself in, although they are liable to be demanding. It is not so much a matter of open questions needing resolution as it is of whole theories needing to be developed. So in a sense the mathematics involved is not like an integral structure to which embellishments may be added, but rather like a partially-formed foundation for an eventually huge and magnificent building. The construction worker would be torn to decide among all the different places he could start adding bricks, or where he could continue the process of building the foundation.

The work which has most heavily influenced the present lectures is that of Sattinger (1973), on which a good part of Chapters I and II is based. Among the other papers which have been most influential, I mention Joseph and Sattinger (1972), Iooss (to appear), Boa (1974), Joseph (1974) Nicolis and Auchmuty (1974), Kopell and Howard (1973), Fife and Greenlee (to appear), Fife (1974).

The reader interested in other general surveys in this and related areas would do well to consult, besides Sattinger (1973), such works as

Iooss (1972-73), Kirchgässner and Kielhofer (1973) (a thorough mathematical account of bifurcation and stability of steady-state solutions of the Navier-Stokes equations), Stakgold (1971), Pimbley (1969), Rabinowitz (1973), Velarde (to appear) (a survey from a fluid dynamicist's point of view of convective bifurcation problems), Sather (1973) (a study of branching of solutions in Hilbert space, especially from multiple eigenvalues, and Hale (1969) (for branching theory of ordinary differential equations).

In presenting a broad survey such as this, the danger is always present of inadvertantly omitting mention of someone's work. I make no claim that the references are anywhere near complete, and they were not intended to be. The reader interested in pursuing a given topic could obtain further references from the bibliographies of the papers cited.

At the time these lectures were envisaged, it was my intention also to outline the present state of branching theory of thin elastic plates and shells. This area of continuum mechanics exhibits very striking branching phenomena. Indeed, the "buckling" of plates and rods can be interpreted as the transition from one solution branch of the relevant equations to another. My intent has had to yield to the present selection of topics, which demand more time and space than were originally allocated to them. A branching theory of the von Karman equations, somewhat in the spirit of the present lectures, was obtained by Berger and Fife (1966, 1968) and Berger (1967). Some important work has been done in recent years on primary and secondary buckling of plates and shells by such researchers as Srubščik, Reiss, Keller, Bauer, Stroebel, Warner, Cheo, and others.

P. C. Fife

I. PRELIMINARIES

1. The Implicit Function Theorem

Most of the results dealing with branching phenomena that I wish to describe in these lectures can be based on a judicious use of the implicit function theorem.* Though it may not appear the same, this approach is the essence of the classical Ljapunov-Schmidt method (see, for example, Vainberg and Trenogin (1962, 1969). For a more explicit use of the implicit function theorem in bifurcation problems, see Crandall and Rabinowitz (1971) and Sattinger (1973).

Our problems involve a solution u and a parameter λ. It will be convenient to express both of them in terms of a third quantity, a "perturbation parameter" ε. In some of our problems, λ may be a vector (even infinite dimensional), rather than just a real number. Accordingly we shall deal with a "solution space" X, of which u is to be an element, and a "parameter space" Λ, containing λ. Besides the Banach spaces X and Λ, we shall deal with another Banach space W, the range space of our operators. We assume W to be decomposable as a direct sum $W = Y \oplus Z$ for some spaces Y and Z. Given $W = y + z \in W$, we define the projections $Pw = y \in Y$, $Qw = (I - P)w = z \in Z$.

In these terms, the following is the form of the implicit function theorem which we shall use:

THEOREM 1. *Let* $F(u,\lambda,\varepsilon):X \times \Lambda \times \mathbb{R} \to W$ *be a mapping, continuously differentiable near* $u = \lambda = \varepsilon = 0$, *satisfying*:

(i) $F(0,0,0) = 0$;

---

*For other approaches, see, for example, the lectures by Rabinowitz at this meeting, and Krasnosel'skii (1964).

(ii) $Q\frac{\partial F}{\partial u}(0,0,0): X \to Z$ *is one-one and onto;*

(iii) $P\frac{\partial F}{\partial \lambda}(0,0,0): \Lambda \to Y$ *is one-one and onto;*

(iv) $P\frac{\partial F}{\partial u}(0,0,0) \equiv 0.$

Then for small enough ε, *there exists a solution* $u(\varepsilon) \in X$, $\lambda(\varepsilon) \in \Lambda$, *of*

$$F(u, \lambda, \varepsilon) = 0, \tag{1.1}$$

which is continuous in ε *and satisfies* $u(0) = 0$, $\lambda(0) = 0$.

Proof. For each ε, F can be thought of as a mapping from $X \times \Lambda$ into $Y \times Z$, with Frechet derivative expressible as a 2×2 Jacobian matrix

$$\frac{\partial F}{\partial(u,\lambda)} = \begin{bmatrix} P\frac{\partial F}{\partial u} & P\frac{\partial F}{\partial \lambda} \\ Q\frac{\partial F}{\partial u} & Q\frac{\partial F}{\partial \lambda} \end{bmatrix}.$$

At $u = \lambda = \varepsilon = 0$, the operator in the position (1,1) in this matrix is identically zero, and those in positions (1,2) and (2,1) are invertible. Finally, $Q\frac{\partial F}{\partial \lambda}$ is bounded. It follows immediately that the whole Jacobian $\frac{\partial F}{\partial(u,\lambda)}(0,0,0)$ is invertible. Hence the conclusion follows by the usual form of the implicit function theorem.

There follow several applications of this theorem.

2. Bifurcation from Simple Eigenvalues

Consider the equation

$$(A + \mu B)U + NU = 0 \tag{2.1}$$

for a vector U in a Banach space $\hat{X}$. Here μ is a real parameter, A and B are linear operators from $\hat{X}$ onto another space W, and N is a twice differentiable higher order operator: $\|NU\|_W = O(\|U\|^2_{\hat{X}})$ as $\|U\|_{\hat{X}} \to 0$. It is assumed that for some value $\mu = \mu_0$, the operator $A_0 \equiv A + \mu_0 B$ has a one-dimensional nullspace spanned by some ϕ, and that its range can be characterized by $R_{A_0} = \{w \in W: \langle \phi^*, w\rangle = 0\} \equiv Z$, for some $\phi^* \in W^*$. Let Q be a projection of W onto Z, $P = I - Q$, and $Y = PW$. Similarly let $\hat{P}$ be a projection of $\hat{X}$ onto the null space of A_0, $\hat{Q} = I_{\hat{X}} - \hat{P}$, and $X = \hat{Q}\hat{X}$. Since $\hat{X} = (\text{nullspace}) \oplus X$, any given vector $U \in \hat{X}$ can be written $U = \varepsilon(\phi + u)$ with the scalar ε and vector $u \in X$ uniquely determined. Finally, let $\lambda = \mu - \mu_0$. Thus, the problem (2.1) can be reformulated in terms of u, λ, and ε:

$$(A_0 + \lambda B)(\varepsilon\phi + \varepsilon u) + N(\varepsilon\phi + \varepsilon u) = 0.$$

Dividing by ε and using the fact that $A_0\phi = 0$, we have

$$F(u,\lambda,\varepsilon) = 0, \tag{2.2}$$

where $F(u,\lambda,\varepsilon) = (A_0 + \lambda B)u + \lambda B\phi + \frac{1}{\varepsilon}N(\varepsilon\phi + \varepsilon u)$. Hypothesis (i) in Theorem 1 is immediate. Since $Q\frac{\partial F}{\partial u}(0,0,0) = QA_0 = A_0$, and A_0 is one-one from X to $R_{A_0} = Z$, we have that (ii) is also true. Since P is a projection onto a one-dimensional subspace Y, and

$\frac{\partial F}{\partial \lambda}(0,0,0)\lambda' = (B\phi)\lambda'$, (iii) reduces to the requirement that $PB\phi \neq 0$ or

$$\langle \phi^*, B\phi \rangle \neq 0. \tag{2.3}$$

It may be shown by a straight-forward argument (see Sattinger (1973)) that (2.3) can be given an alternate form: For each μ, let $\gamma(\mu)$ be an eigenvalue of $A + \mu B$. In particular, let it denote the continuous eigenvalue function satisfying $\gamma(\mu_0) = 0$. Assume γ is continuously differentiable. Then (2.3) may be replaced by

$$\gamma'(\mu_0) \neq 0. \tag{2.4}$$

Finally, (iv) is true by definition of P. We apply Theorem 1 to obtain

THEOREM 2. *If (2.3) (or (2.4)) holds, then under the above assumptions, there exist functions* $U(\varepsilon)$, $\mu(\varepsilon)$, *defined for small* ε, *satisfying* $U(0) = 0$, $\mu(0) = \mu_0$, $\hat{P}U(\varepsilon) = \varepsilon\phi$, *and* (2.1).

This is a "bifurcation" theorem for the following reason. The trivial function $U = 0$ is a solution of (2.1) for all values of μ. The conclusion of Theorem 2, on the other hand, is that there exist nontrivial solutions as well, for certain values of μ near μ_0, namely, for the set of values of the function $\mu(\varepsilon)$. We know these solutions are nontrivial for $\varepsilon \neq 0$, because of the condition $\hat{P}U(\varepsilon) = \varepsilon\phi$. Thus ε serves as a parameter for a continuous branch of nontrivial solutions $U(\varepsilon)$ intersecting the trivial branch at $\varepsilon = 0$, $\mu = \mu_0$. Actually, one could speak of two branches: that with $\varepsilon \geq 0$, and one with $\varepsilon \leq 0$.

Bifurcation from known nonzero families of solutions may be handled this same way by reformulating the problem in terms of the difference between the desired bifurcated solution and the known one. Calling this difference U, we obtain a problem of the type treated in this section.

The sign of $\mu(\varepsilon) - \mu_0$ for the various branches is important for stability considerations. Suppose, for the sake of definiteness, that when $\mu < \mu_0$, all the eigenvalues of $A + \mu B$ have positive real part, and that the "critical" eigenvalue, $\gamma(\mu)$, namely the one which vanishes when $\mu = \mu_0$, passes into the left half plane when $\mu > \mu_0$. The "linear stability criterion" for the zero solution with respect to the evolution problem (3.1) below with time derivative added is simply the criterion that all eigenvalues of $A + \mu B$ have positive real parts. Thus, the situation we are describing is when the zero solution loses stability as μ increases through the value μ_0. Then a bifurcation branch for which $\mu(\varepsilon) \geq \mu_0$ is called a supercritical branch, and one for which $\mu(\varepsilon) \leq \mu_0$ is called subcritical.

It has been shown fairly generally that nontrivial solutions on supercritical branches are stable for small ε, and those on subcritical ones are unstable. For various cases of this, see Judovič (1967), Kirchgässner and Sorger (1968), and Sattinger (1971). For more precise results see Crandall and Rabinowitz (to appear).

3. Bifurcation of Periodic Solutions from Steady Ones

We now consider problems in which an evolution variable t appears. Specifically, we add a t-derivative to (2.1), obtaining

$$\frac{\partial U}{\partial t} + (A + \mu B)U + NU = 0. \tag{3.1}$$

In the case when this represents a system of ordinary differential equations, the basic result of periodic bifurcation was obtained by Hopf (1942).

Typically, A will be a differential operator in space variables, so that U will be a function of t and other variables. To emphasize its t-dependence, we sometimes write $U(t,\cdot)$. The operators A, B, and N are supposed to have properties similar to those outlined in Section 2 above, but, of course, they act on a space in which the new operator $\frac{\partial}{\partial t}$ has meaning. More will be said later about this space. We shall seek solutions U which are periodic in time with frequency ω.

As in the preceding section, the problem always has a trivial solution $U \equiv 0$. Our object is to give a sufficient condition for the existence of a bifurcating nontrivial branch.

Since ω will not be fixed, but will rather depend on the other parameters in the problem, it is convenient to normalize the problem somewhat by introducing a new time variable $s = \omega t$, and to consider U to be a function of s, rather than of t. Then we seek solutions $U(s,\cdot)$ which are periodic with period 1. Furthermore we define a function V by $U(s - \delta,\cdot) = \varepsilon V(s,\cdot)$, where ε and δ are two real numbers to be specified later. In all, (3.1) becomes

$$\omega\frac{\partial V}{\partial s} + (A + \mu B)V + \varepsilon\hat{N}(V,\varepsilon) = 0, \tag{3.2}$$

where $\hat{N}(V,\varepsilon) \equiv \frac{1}{\varepsilon^2}N(\varepsilon V)$.

The problem is to find periodic solutions of (3.2), with period 1, which are nontrivial (since $V = 0$ is always a solution). But we must be more precise about the spaces we are working in, and about the assumptions on the operators. First consider, for simplicity, the case when U is an n-vector function of s and x, defined for $0 \leq x \leq 1$, and

$A = \frac{-d^2}{dx^2} + E$, E being an $n \times n$ matrix. The domain of A is restricted to functions $U(s,x)$ satisfying $U(s,0) = U(s,1) = 0$. Also suppose that B is a matrix. As in Section 2, we assume that for some value $\mu = \mu_0$, the spectrum of the operator $A_0 = A + \mu_0 B$ has a special property. Whereas in Section 2 this property consisted of A_0 having a simple eigenvalue at the origin, we now assume that there is a simple eigenvalue at $i\omega_0$ for some $\omega_0 > 0$. Moreover, we assume that A_0 has no other eigenvalues of the form $in\omega_0$, for n a nonnegative integer. Since $A_0\phi = \overline{A_0\overline{\phi}}$,* it follows that there exists a second simple eigenvalue at $-i\omega_0$. Let ξ and $\overline{\xi}$ be the corresponding eigenvectors, so that $A_0\xi = i\omega_0\xi$ and $A_0\overline{\xi} = -i\omega_0\overline{\xi}$. Similarly, let η, $\overline{\eta}$ be eigenvectors of $A_0^* = -\frac{d^2}{dx^2} + E^* + \mu_0 B^*$: $A_0^*\eta = -i\eta$, $A_0^*\overline{\eta} = i\overline{\eta}$. Then (by normalization) we can guarantee that $(\xi,\eta)_0 = 1$ and $(\xi,\overline{\eta})_0 = 0$ in the sense of the scalar product $(\cdot,\cdot)_0$ on $L_2(0,1)^{(n)}$ (complex).

Now let $H^c = L_2((0,1) \times (0,1))^{(n)}$ (complex), restricted to functions $U(s,x)$ satisfying $U(0,x) = U(1,x)$ (periodicity conditions). Also let H^r denote the analogous real Hilbert space. The operators A and B are still defined on a dense subset of these spaces H^c and H^r.

Let $J = \omega_0 \frac{\partial}{\partial s} + A_0$. We shall find the nullvectors of J. If ψ is one, then we may expand in Fourier series: $\psi = \sum_n \chi_n(x)e^{ins}$, and have

$$0 = J\psi = \sum(in\omega_0 I + A_0)\chi_n e^{ins},$$

which implies that for each n, either $\chi_n = 0$ or $-in\omega_0$ is an eigenvalue of A_0. The only eigenvalues of A_0 within the set $\{in\omega_0\}$ are $\pm i\omega_0$. Hence the only nullvectors of J are $\phi \equiv \chi_1 e^{is} = \xi e^{is}$, and

*Superbars denote complex conjugation.

$\chi_{-1}e^{-is} = \bar{\xi}e^{-is} = \bar{\phi}$. Similarly, the only nullvectors of J^* are $\phi^* = e^{-is}\eta$ and $\bar{\phi}^*$.

The Fredholm alternative holds for J, so that $Ju = f \in H^c$ has a solution if and only if f is orthogonal to all nullvectors of J^*: $(f,\phi^*) = (f,\bar{\phi}^*) = 0$. In the case that f is real, these two conditions can be written as one: $(f,\phi^*) = 0$. Let $[f]$ be the complex-valued functional $[f] \equiv (f,\phi^*)$. We therefore have that the range of J, restricted to real vectors in its domain, is the set $\{f \in H^r : [f] = 0\}$.

It is easily verified that $[\phi] = 1$, $\left[\bar{\phi}\right] = 0$, $[\frac{\partial\phi}{\partial s}] = -i$, $[\frac{\partial\bar{\phi}}{\partial s}] = 0$. Using these facts, it can be checked that the operator

$$Pu \equiv [u]\phi + \overline{[u]}\bar{\phi} \equiv 2\mathrm{Re}([u]\phi) \tag{3.3}$$

is a projection onto the nullspace of J. Finally, since ϕ and $\bar{\phi}$ are linearly independent, it follows that $Pu = 0$ if and only if $[u] = 0$. It further follows from this and the above characterization of the range of J, restricted to real functions, that this range is $(I - P)H^r$.

In summary, we have the following property, which we call

PROPERTY $\mathcal{P}(A,B,H^r)$. <u>If</u> $A_0 = A + \mu_0 B$ <u>has</u> $\pm i\omega_0 (\omega_0 > 0)$ <u>as simple eigenvalues, and no others of the form</u> $in\omega_0$ (<u>n an integer</u>), <u>then</u>

(1) J <u>has a two-dimensional nullspace spanned by some complex vectors</u> ϕ <u>and</u> $\bar{\phi}$;

(2) <u>there exist projections</u> P <u>and</u> $Q = I - P$ <u>onto the nullspace and range of</u> J <u>respectively</u>;

(3) <u>there exists a complex-valued linear functional</u> $[u]$ <u>on</u> H <u>such that</u>

(a) $Pu = 2\mathrm{Re}([u]\phi)$,

(b) $[\frac{\partial\chi}{\partial s}] = -i[\chi]$ <u>for nullvectors</u> χ <u>of</u> J; <u>and</u>

(c) $[U(s - \delta,\cdot)] = e^{-i\delta}[U(s,\cdot)]$.

(Incidentally, this last property (c) may be verified from the definitions of $[\cdot]$ and ϕ^*.)

Of course, the property P holds far more generally than in the case given. We now go back to the original problem (3.2), and put it into a general setting in which the property still has meaning.

Let $\tilde{\hat{X}}$ and $\tilde{\hat{W}}$ be two real Banach spaces whose elements are functions of s and possibly other (space) variables, and are periodic in s with period 1. We assume that $\tilde{\hat{X}} \subset \tilde{\hat{W}}$, that $\partial/\partial s$, A, B, and $\hat{N}$ are bounded (in the case of $\hat{N}$, uniformly for small ε) operators from $\tilde{\hat{X}}$ into $\tilde{\hat{W}}$, and that $\hat{N}$ is twice continuously differentiable near $V = \varepsilon = 0$. The property P defined above can be interpreted in this new setting; to do so, one requires the projections P and Q and the linear functional $[\cdot]$ to act on the space $\tilde{\hat{W}}$; by inclusion, they will then also be defined on $\tilde{\hat{X}}$. We denote the property by $P(A,B,\tilde{\hat{X}},\tilde{\hat{W}})$. It may or may not hold. Joseph and Sattinger (1972) have shown that it does hold when A_0 is a linear Navier-Stokes operator (to be used later), and for $\tilde{\hat{X}}$ and $\tilde{\hat{W}}$ appropriate Hölder spaces (for example). Furthermore, Sattinger (1973) has shown that it holds for more general operators A and B.

We assume $P(A,B,\tilde{\hat{X}},\tilde{\hat{W}})$ to hold. Returning to the definition of V, we now specify the real parameters ε and δ in terms of the sought solution U by the formula $[U] = \varepsilon e^{i\delta}$.

It follows from the definition of V and from part 3(b) of Property $\mathcal{P}$ that $[V] = 1$, hence that $PV = 2\mathrm{Re}([V]\phi) = 2\mathrm{Re}\phi \equiv q$. We may thus decompose the solution V in the form $V = PV + QV = q + u$, where $Pu = [u] = 0$, so that $u \in Q\tilde{X}$, which space we denote by X. Using these expressions and defining λ to be the vector $\lambda = (\omega - \omega_0, \mu - \mu_0)$, we rewrite (3.2) in the form

$$F(u,\lambda,\varepsilon) \equiv Ju + (\omega - \omega_0)\left(\frac{\partial q}{\partial s} + \frac{\partial u}{\partial s}\right)$$

$$+ (\mu - \mu_0)B(q + u) + \varepsilon\overline{N}(u,\varepsilon) = 0, \tag{3.4}$$

where $\overline{N}(u,\varepsilon) \equiv \hat{N}(q + u,\varepsilon)$.

The range of F is in $\tilde{W}$, which we identify with the space W in Theorem 1. Finally defining $Y = P\tilde{W}$, $Z = Q\tilde{W}$, we have $\tilde{W} = Y \oplus Z$.

The problem has been put into the framework of Theorem 1, and we need to verify the assumptions of that theorem.

(i) Immediate.

(ii) $Q\frac{\partial F}{\partial u}(0,0,0) = QJ = J$. According to part (2) of Property $\mathcal{P}$ and the definitions of X and Z, this is an invertible operator from X onto Z.

(iii) Given any $p \in Y = P\tilde{W} = P\tilde{X}$, we must be able to solve

$$P\frac{\partial F}{\partial \lambda}(0,0,0)\tilde{\lambda} = p \tag{3.5}$$

uniquely for $\tilde{\lambda} = (\tilde{\omega} - \omega_0, \tilde{\mu} - \mu_0) \in \Lambda$. But p, being real and a linear combination of ϕ and $\overline{\phi}$, can be written in the form

$p = \alpha\phi + \overline{\alpha\phi}$ for some complex number α. Using (3.3), we see that (3.5) becomes

$$[\frac{\partial F}{\partial \lambda}\tilde{\lambda}]\phi + \overline{[\frac{\partial F}{\partial \lambda}\tilde{\lambda}]\phi} = \alpha\phi + \overline{\alpha\phi},$$

which is equivalent to $[\frac{\partial F}{\partial \lambda}\tilde{\lambda}] = \alpha$. But $[\frac{\partial F}{\partial \lambda}\tilde{\lambda}] = (\tilde{\omega} - \omega_0)[\frac{\partial q}{\partial s}] + (\tilde{\mu} - \mu_0)[B\phi] = -i(\tilde{\omega} - \omega_0) + (\tilde{\mu} - \mu_0)[B\phi]$, so we have

$$-i(\tilde{\omega} - \omega_0) + (\tilde{\mu} - \mu_0)[B\phi] = \alpha.$$

This is solvable for $\tilde{\omega} - \omega_0$ and $\tilde{\mu} - \mu_0$ if and only if

$$\text{Re}[B\phi] \neq 0, \tag{3.6}$$

which we hereby assume.

Again, it was shown by Sattinger that (3.6) can be replaced by the condition that

$$\text{Re}\gamma'(\mu_0) \neq 0, \tag{3.7}$$

where $\gamma(\mu)$ is the eigenvalue of $A + \mu B$ satisfying $\gamma(\mu_0) = i\omega_0$.

(iv) We have $P\frac{\partial F}{\partial u}(0,0,0) = PJ = 0$, since $P = I - Q$.

In all, the assumptions (i), (ii), and (iv) of Theorem 1 are verified, and (iii) assumed. We conclude:

THEOREM 3. *Assume* $\pm i\omega_0$ *are simple eigenvalues of* A_0, *that there are no others of the form* $in\omega_0$, *and that* $P(A,B,\tilde{X},\tilde{W})$ *and* (3.6) *hold. Then there exist, for small enough* ε, *functions* $V(\varepsilon) \in X$, $\mu(\varepsilon)$, *and* $\omega(\varepsilon)$ *satisfying* (3.2), $[V] = 1$, $\omega(0) = \omega_0$, *and* $\mu(0) = \mu_0$.

This yields a solution $U_\varepsilon(t,\cdot) = \varepsilon V_\varepsilon(\omega(\varepsilon)t,\cdot)$ of the (incompletely formulated) problem (3.1), periodic with frequence $\omega(\varepsilon)$. It, of course, represents a nontrivial branch of solutions.

As in the last section, one can speak of sub- and super-critical branches. Suppose the spectrum of A_μ is in the right half plane for $\mu < \mu_0$, and that $\mathrm{Re}\lambda'(\mu_0) < 0$. Then branches with $\mu(\varepsilon) > \mu_0$ are supercritical, and those with $\mu(\varepsilon) < \mu_0$ are subcritical. It is an important fact, shown by Joseph and Sattinger (1972), that (in the case of the Navier-Stokes equations, and probably quite generally), solutions on supercritical branches for small ε are stable, whereas those on subcritical ones are unstable.

4. An Example with Λ Infinite Dimensional*

We consider the problem of finding 2π-periodic solutions of the nonlinear wave equation

$$\Box U = U_{tt} - U_{xx} = \varepsilon G[U], \qquad 0 \leq x \leq \pi, \tag{4.1}$$

$$U(0,t) = U(\pi,t) = 0 \tag{4.2}$$

where G is some smooth function of x, t, u, and its first derivatives, or perhaps some nonlocal differentiable operator acting on an appropriate space (see below) of functions of x, t, 2π-periodic in t.

---

*Another such example will be discussed in Section III.4.

One of the first serious studies of this type of problem was made by Vejvoda (1964), and there has been much work done on it by many authors since that time. Rather than report on this work, I shall show how one may obtain some naive preliminary results by fitting the problem into the foregoing framework.

Let H_k denote the usual Sobolev Hilbert spaces, on the rectangle $(0,\pi) \times (0,2\pi)$, but restricted to functions satisfying periodic boundary conditions in t. Let $\overset{\circ}{H}_k$ be the subspace satisfying the boundary conditions (4.2). For some $k \geq 2$, let Λ be the subspace of functions in $\overset{\circ}{H}_k$ satisfying $\square v = 0$, and $X = H_k \cap \Lambda^{\perp}$, $\Lambda^{\perp}$ being the orthogonal complement of Λ in H_0. Let P denote orthogonal (in H_0) projection of $\overset{\circ}{H}_k$ onto Λ. It was proved by Rabinowitz (1967) that $\square^{-1}$ exists as a compact mapping from X into itself.

We assume that there exists a function $V^0 \in \Lambda$ satisfying $PG[V^0] = 0$ (giving verifiable conditions on G under which this assumption is valid has been a major task in the past). Thus $G[V^0] \in X$, and we may define $U^1 = \square^{-1}G[V^0]$. Since $H_k = X \oplus \Lambda$, we may decompose the desired solution U in the form

$$U = V^0 + v + \varepsilon(U^1 + u), \tag{4.3}$$

where $v \in \Lambda$ and $u \in X$. Substituting (4.3) into (4.1), we obtain

$$\square u = K[\varepsilon u + v] \equiv G[V^0 + v + \varepsilon(U^1 + u)] - G[V^0]. \tag{4.4}$$

Applying the projection P and the operator $\square^{-1}(I - P)$ successively to (4.4), we write it in the equivalent form

$$F(u,v,\varepsilon) = 0, \tag{4.5}$$

where F is the sum of two terms,

$$F(u,v,\varepsilon) \equiv PK[\varepsilon u \pm v] + (u - \square^{-1}(I - P)K[\varepsilon u + v]) \in \Lambda \oplus X \equiv W \equiv H_k.$$

Since the two terms are in complementary subspaces, (4.5) is equivalent to their vanishing individually. Furthermore, notice that any solution (u,v) of (4.5) automatically satisfies the boundary condition (4.2); in fact, v does by definition of Λ, and u does because $u - \square^{-1}(I - P)K = 0$, implying that u is in the range of $\square^{-1}$.

In the context of Theorem 1, we define $Y = \Lambda$ and $Z = X$. Let us examine the assumptions of that theorem.

(i) Immediate.

(ii) $Q\frac{\partial F}{\partial u} = I_X - \square^{-1}(I - P)\ \varepsilon K'[\varepsilon u + v]$, so $Q\frac{\partial F}{\partial u}(0,0,0) = I_X$, which is invertible.

(iii) $P\frac{\partial F}{\partial v}(0,0,0) = PK'[0]|\Lambda = PG'[V^0]|\Lambda$.

We <u>assume</u> this is an invertible mapping from Λ onto Λ.

(iv) $P\frac{\partial F}{\partial u} = \varepsilon PK'[\varepsilon u + v]$, so $P\frac{\partial F}{\partial u}(0,0,0) = 0$.

Therefore we have

THEOREM 4. <u>Under the above assumptions, for small</u> ε <u>there exists a periodic function</u> $U(\varepsilon) \in H_k$ <u>of the form</u> (4.3), <u>satisfying</u> $U(0) = V^0$, (4.1), and (4.2).

P. C. Fife

II. FLUID DYNAMICS

1. Introduction

Fluid dynamics presents one of the richest areas where branching phenomena can be seen. Consider a thought-experiment as follows. Stir a cup of tea in a regular manner. If the stirring proceeds very slowly, then what you might expect is a fairly regular periodic motion of the fluid, at least after the stirring has proceeded for a long enough time. If the stirring rate increases, however, peculiar things might start to happen. Vortices may start to form near the stirring rod, and dissipate or become amalgamated with the general confusion of the fluid motion, which grows increasingly more and more complex. Finally, when the stirring is fast, the fluid motion appears to have very few regular properties and is described by the all-inclusive term "turbulent". In this experiment one can concoct a parameter, the Reynolds number, which is some measure of the rate of stirring. As the Reynolds number increases, the resulting motion becomes more and more complex, and new patterns of motion develop.

It has been conjectured that new kinds of patterns arise as the Reynolds number passes through certain discrete values, which may or may not be bifurcation points. Take a much simpler situation -- that of two concentric cylinders, the inner one rotating (Taylor (1923)). This would be a simple type of stirring. Then the fluid motion for very small Reynolds numbers is of a very simple character: the fluid particles move in concentric circles, and their velocity is a very simple function of radius only. This is called Couette flow. Furthermore, this is the only flow possible. However if the Reynolds number R is increased past the first of this discrete set of points, the so-called first critical value, then a second type of flow appears: Taylor vortex flow. This type of motion also has the property that the fluid velocity is independent of time; but the particles no longer move in concentric circles, but rather in helices wrapped around the inner cylinder.

Most people assume that the incompressible Navier-Stokes equations provide an adequate model for this situation; and practically all of our discussion will be within that framework. The first fact to recognize is that there is indeed an exact solution to these equations corresponding to Couette flow, as described above. In fact, this is true for arbitrary Reynolds number, but the flow is not seen in practice for large R. This illustrates the point that even the wealth of actual motions seen in nature in a situation like this gives us only a part of the possible solutions of the Navier-Stokes equations; some are simply not seen. The question of which <u>are</u> observed is tied up with stability considerations.

In the cylinder problem posed above, we now ask what kind of motion appears at the next critical point. It is thought, but not proved, that whereas the first two kinds of motion were steady, there now appears a motion which is periodic in time. And at the next critical point, it is thought, or at least the possibility is held open, that more complex periodic motions or quasiperiodic motions become possible and are seen. And after that, who knows what?

The types of flow I have mentioned so far can be described in terms of certain invariant sets. In the mathematical analysis of the Navier-Stokes equations, one often thinks of a solution as being a function $u(t)$ of time with values in some suitably chosen Banach Space C, so

$u(t) \in C$ for all t. The elements of C represent vector-valued functions of the space variables, the components of the vector being the three velocity components of the fluid, and the pressure. So let $u(t)$ denote a solution. The first two kinds of motion which appear in the Taylor cylinder problem are steady motions; i.e., u does not depend on t. Thus they are merely points in C. To say that they are legitimate solutions of the N-S equations is equivalent to saying that if one of them is used as initial value in the evolution problem, then the solution of that problem is constant. In other words, such a point would be an *invariant set* with respect to the evolution problem: if the system is ever at that point, it remains there.

For the problems we are going to speak about, there is a well-defined solution operator $S(t,u_0)$ $(t > 0)$ which maps elements $u_0 \in C$ (initial values) into the corresponding values $u(t)$ of the solution at time t. Steady state solutions, then, are elements u_0 which are left invariant by $S(t,u_0)$ for all $t > 0$.

In the same way, a periodic solution has an invariant set which is a closed one-dimensional loop in C. Thus, if u_0 is on the loop, then so is $S(t,u_0)$ for all $t > 0$. Finally, certain motions (it is conjectured) might have other kinds of finite-dimensional invariant manifolds, but not be periodic. For example, the appearance of invariant two-dimensonal tori is a possibility, as we shall see. Presumably, for higher Reynolds numbers, motions with invariant sets of higher (but finite) dimension may exist. Ladyženskaja (1972) has indicated the proper definition of such invariant sets, and has shown that the Navier-Stokes problem for flow in a bounded domain with forcing term, and zero boundary conditions, when restricted to such a set, defines a dynamical system. She furthermore expresses the opinion that the dynamical system is the "central object" of study in the investigation of such evolution problems. The study of how the structure of these invariant sets changes with Reynolds number would be a tremendously difficult but important undertaking. To indicate how difficult it is, let me say that to this point, it has only progressed one or two steps, and for only a few special situations. This is in spite of the large amount of sophisticated mathematical thought which has been applied to the Navier-Stokes equations.

The actual transition from one type of invariant set to another might be accomplished by means of bifurcation phenomena, as was alluded to above. An idea similar to this was expressed by Landau (1944), who suggested that the transition to turbulence is by repeated branching into more complicated quasiperiodic motions. On the other hand, Joseph and Sattinger (1972) discuss this point and bring out that transitions are often made by "snapping through", in which the system makes a sudden finite transition from one mode to another. Transitions of the "snap-through" type occur through subcritical branches, and would therefore seem to be excluded when all branching is supercritical.

Models, simpler than the theory of the Navier-Stokes equations, have been given to illustrate or exhibit repeated branching of the Landau type. One was given by Hopf (1958). A more realistic branching theory (Joseph (1974)) is discussed below in Section 6.

Regarding the question of which fluid motions can actually be seen in nature, one must recognize that some invariant sets are stable and others not. Physically observable motions are not seen on the second kind of set. Stability considerations are a very important part of the study of branching of solutions in fluid dynamics; yet I will say relatively little about this aspect in my lectures, due to time limitation.

Finally, the obvious fact should be emphasized that transient motion, as well as motion on a stable invariant set, occurs. But such motion is not the object of our study. In this connection, it is possible (see, for example, Joseph (to appear) for a discussion), that turbulent motion actually consists to a large extent of transient flows passing from one of a large number of possible stable invariant sets to another.

With all of this said, it is time to delineate the subject matter of my lectures in fluid dynamics. My main purpose will be to describe the mathematical aspects of some things which are known about transitions between the simpler types of invariant sets by means of bifurcation.

2. Formulation of the Evolution Problem

The Navier-Stokes equations may be written in the form

$$\frac{\partial U}{\partial t} + \mu U\cdot\nabla U + \nabla p = \Delta U + f(x,t) \tag{2.1a}$$

$$\nabla\cdot U = 0, \tag{2.1b}$$

where $U = (U_1,U_2,U_3)$ is the velocity field and p is the pressure. We shall consider U and p to be defined for x in the closure $\overline{\Omega}$ of a smooth bound domain, and for $t \geq 0$. On the boundary $\partial\Omega$ we prescribe the velocity

$$U|_{\partial\Omega} = H(x,t). \tag{2.2}$$

(H must satisfy the condition $\int_{\partial\Omega} H \cdot n dA = 0$ for consistency with (2.1b).)

Other variants of this problem, such as with other types of boundary conditions, may also be treated. The parameter μ above is the Reynolds number, a dimensionless constant expressible as a certain combination of various characteristic dimensions of the problem and of the coefficient of viscosity.

To put the system into a more mathematical tractable form, one first decides on a Banach space $\mathcal{B}$ to be the domain of the solutions. There are several possible choices for $\mathcal{B}$. See, for example, Ladyženskaja (1963). For the sake of definiteness, we shall take the velocity U to be an element of the space $C^{(3)}_{2+2\alpha,1+\alpha}(\overline{D})$, $0 < \alpha < \frac{1}{2}$, where $D = \Omega \times (0,T_1)$ for some positive number T_1. This means that each of the three components of U has x-derivatives to order two and one t-derivative, all of them Hölder continuous. The Hölder exponent is 2α in the variable x, and α in t. More particularly, we restrict U to be the subspace of divergence-free (solenoidal) vector fields with these properties. The function p is viewed as an element in a similar space. Let P denote the projection of $C^{(3)}_1(\Omega)$ onto its solenoidal subspace, self-adjoint with respect to the L_2 norm. It may be extended to be operable even on spaces, such as we shall encounter, where elements are not necessarily differentiable. If we now apply P to (2.1a), we eliminate the function p from consideration (and, of course, (2.1b) is superfluous). So doing, we obtain an equation of the form

$$\frac{\partial U}{\partial t} + A_0 U + \mu N U + F = 0, \tag{2.3}$$

where $A_0 = -P\Delta$, $NU = PU\cdot\nabla u$, and $F = -Pf$. It turns out that the problem (2.3) is equivalent to (2.1), in the sense that every solution U of (2.3) determines a function $p(x,t)$, unique except for an arbitrary additive function of t, so that (2.1) is satisfied.

In many of the problems we are going to mention, we shall suppose known a particular family of solutions $U_\mu^0(x,t)$ of (2.3), (2.2), analytic in μ for μ in some interval $I \subseteq \mathbb{R}$, and we shall seek other solutions in the form

$$U = U_\mu^0 + u. \tag{2.4}$$

With this substitution, (2.3) becomes

$$\frac{\partial u}{\partial t} + A_0 u + \mu B_\xi u + \mu N u = 0, \tag{2.5}$$

where $B_\mu u = PU_\mu^0 \cdot \nabla u + Pu \cdot \nabla U_\mu^0$. Since u must vanish on $\partial\Omega$, we further restrict to a subspace of functions so vanishing. Thus in all, our solutions u will be considered elements of the space $\mathcal{B}$, defined as consisting of functions in $PC^{(3)}_{2+2\alpha,1+\alpha}(\overline{D})$ which vanish on $\partial\Omega$ for all $t \in [0,T_1]$.

We also define $\mathcal{B}_0 = PC^{(3)}_{2\alpha,\alpha}(\overline{D})$, and $\mathcal{C}$ as the space of divergence free vector functions $u(x) \in C^{(3)}_{2+2\alpha}(\Omega)$ vanishing on $\partial\Omega$. Finally, $\mathcal{C}_0 = PC^{(3)}_{2\alpha}(\Omega)$.

The following properties of the operators involved in (2.5) will be important in what follows:

1. For each $\mu \in I$, the operator $A_\mu \equiv A_0 + \mu B_\mu$, considered as

an operator from $\mathcal{C}$ into $\mathcal{C}_0$, has nullspace with finite dimension equal to the codimension of its range.

2. Let $T_1 = \infty$, and let $\tilde{\mathcal{B}}$ denote the subspace of $\mathcal{B}$ consisting of functions periodic in time with period (say T. Let $\tilde{\mathcal{B}}_0$ be the similar subspace of $\mathcal{B}_0$. Let $J_\mu = \frac{\partial}{\partial t} + A_\mu$. Then for each $\mu \in \mathcal{I}$, J_μ obeys the Fredholm alternative. In fact, Property $\mathcal{P}(A, B_\lambda, \tilde{\mathcal{B}}, \tilde{\mathcal{B}}_0)$ (defined in Section I.3) holds. (See Joseph and Sattinger (1972)).

As mentioned, this evolution problem can be formulated within the framework of other Banach spaces as well. If, in the above treatment, we replace $\mathcal{B}$ by $\hat{\mathcal{B}} = C^0(0,T_1;\mathcal{D}) \cap C^1(0,T_1;\mathcal{H})$, where $\mathcal{H} = P(L^2(\Omega))^{(3)}$, $\mathcal{D} = \{U \in P(H^2(\Omega))^3, U|_{\partial\Omega} = 0\}$, and also replace $\mathcal{C}$ by $\mathcal{D}$, $\mathcal{C}_0$ by $\mathcal{H}$, then properties similar to the above still hold true. Iooss (to appear) has shown the following property to hold; it no doubt holds within the context of $\mathcal{B}$ as well:

3. For each $T_1 > 0$, there exists a neighborhood $\mathcal{N}$ of the origin in $\mathcal{D}$ such that for each $u_0 \in \mathcal{N}$, there exists a solution $u(x,t)$ of (2.5) in $\hat{\mathcal{B}}$ satisfying $u(x,0) = u_0(x)$. The corresponding solution operator $S_{\lambda,T_1}: u_0 \to u(\cdot,T_1)$ is analytic in $\mathcal{D}$ and has compact derivative.

3. Bifurcation of Steady-State Solutions

Suppose the given family of solutions U^0 is time-independent; we ask now whether other time-independent solutions of (2.4) can exist for some μ. Equation (2.5) reduces to

$$A_0 u + \mu B_\mu u + \mu N u = 0. \tag{3.1}$$

It is known that for small enough μ, (3.1) has only the trivial solution $u = 0$; we look for nonzero solutions for larger values of μ. Let us suppose that for some $\mu_0 \in I$, the linear operator $A_{\mu_0} \equiv A_0 + \mu_0 B_{\mu_0}$ has zero as a simple eigenvalue. Then this would be a case treated by Theorem 2 on Page 8, except for the fact that now B_μ and the last term in (3.1) depend on μ. But this is a minor problem, and the conclusion of that example still holds:

If 0 is a simple eigenvalue of A_{μ_0}, and $PB_{\mu_0}\phi \neq 0$, where P is some projection onto the nullspace of A_{μ_0}, and where ϕ is the nullvector:

$$A_{\mu_0}\phi = 0, \tag{3.2}$$

then for small enough ε there exist functions $u_\varepsilon \in C$ and μ_ε such that $u_\varepsilon = \varepsilon(\phi + v_\varepsilon)$, $Pv_\varepsilon = 0$, and (3.1) is satisfied for each ε.

Thus, we have the existence of nontrivial steady-states, provided that certain facts about the spectrum of the linear operator are known. Verifying these facts is unfortunately difficult in general.

For the Taylor rotating cylinder problem, progress has been made. Let me again describe the situation briefly. Consider two coaxial cylinders filled with a viscous incompressible fluid being driven by the inner cylinder, which rotates. For simplicity we assume the outer cylinder is fixed. There is a basic flow, called the Couette flow U^0, which exists for all μ, of the form $U^0_r = U^0_z = 0$; $U^0_\theta = ar + b/r$, where r, θ, z are cylindrical coordinates, subscripts on U represent velocity components in these coordinate directions, and where a and b are constants independent of the viscosity. Thus we may consider U^0 to be independent of μ, if everything but viscosity is held fixed. As in the last section, we seek a flow in the form $U^0 + u$.

The character of the first bifurcation point has not been rigorously established. However, the problem is made more tractable if we arbitrarily delimit the variety of flows by imposing the condition that the flow be axisymmetric, periodic in the z direction, say with given period $2\pi/\sigma$, and be even. Then we need only study the spectrum of the linear operator A_μ restricted to this class of flows. By means of separation of variables, this procedure reduces to an eigenvalue problem for an ordinary differential equation (the independent variable being r). Judovič (1966) has shown that for almost every value of σ, this problem has all of its eigenvalues simple and real. Hence, by the above theory, there certainly exist nontrivial solutions near each of these simple eigenvalues.

Since σ is essentially arbitrary, the above yields a plethora of steady solutions. Not that many are actually observed, of course. For one thing, there is probably a preferred value of σ; and for another, there is a preferred bifurcation point (namely, the first one encountered when μ is increased). All the other solutions will be unstable, because near their bifurcation points A_μ has eigenvalues in the left half-plane. For results on preferred solutions, see Kirchgässner and Sorger (1969), and Kirchgässner and Kielhöfer (1973).

Another problem which has been studied successfully from the point of view of bifurcation theory is the Bénard convection problem (Bénard (1900)), more properly called the Rayleigh convection problem, as

Velarde (1974) has pointed out. This is the situation when a slightly compressible viscous fluid is confined between two horizontal planes held at constant temperatures, the lower plane being the warmer. We assume the "Boussinesq" approximation to hold. This approximation leads to a system of partial differential equations very similar to (2.1), but involving the temperature T as well as the velocity and pressure of the fluid. The boundary conditions are that the velocity must vanish on the two planes, and the temperature T of the fluid is equal to the given constant temperatures on the two planes. There are a variety of parameters appearing in the problem; but if one nondimensionalizes properly, it can be written in terms of only one parameter, the Rayleigh number, which we again denote by μ. This is a dimensionless combination of the viscosity, coefficient of expansion, heat capacity and conduction coefficient, the temperature difference and distance between planes, and the acceleration due to gravity.

There is a very easy time-independent solution to this boundary value problem: the velocity identically zero, and the temperature a linear function of the vertical coordinate. We call this solution the "conduction" solution, and denote it by U^0. Here, of course, U (as well as u below) will have four components instead of three, since temperature must be accounted for. Seeking other steady solutions in the form $U = U^0 + u$, we arrive at an equation much like (3.1):

$$A_0 u + \mu B u + M u = 0, \tag{3.3}$$

where now $A_0 = -P\Delta + E$, E and B are constant matrices, and P is projection onto the subspace of 4-vectors whose first three components (representing the velocity) are divergence-free; and M is quadratic.

As before, the problem is only made tractable by restricting attention to a subclass of possible solutions, defined by certain periodicity and symmetry properties. For its bifurcation analysis, see, for example, Velte (1964) (special case), Judovič (1966), Rabinowitz (1968). So we assume u to be doubly spacially periodic in two horizontal directions, with given periods ℓ_1 and ℓ_2 (one of which could be ∞).

Now if μ_0 is such that the operator $A_0 + \mu_0 B$, restricted to this class, has a simple eigenvalue at the origin, with eigenvector ϕ, then we know that it is a bifurcation point, provided that $PB\phi \neq 0$. This latter condition turns out to be true if $(B\phi,\phi) \neq 0$ in the sense of the L_2 scalar product over the period cell. But since $A_0\phi + \mu_0 B\phi = 0$, the scalar product can be written as $\frac{1}{\mu_0}(A_0\phi,\phi)$. But it can be checked from the form of the matrix E that this can only happen if $\phi = 0$, which is not true.

The only question remaining is whether there is a μ_0 with this property. Again, this linear eigenvalue problem is made simpler by restricting further the class of solutions to those with certain symmetry properties. With the ℓ_i fixed and these symmetry conditions imposed upon u, the linear eigenvalue problem may be analyzed with Fourier series techniques. It turns out that the problem does indeed have lots of simple eigenvalues. (Actually this is not always true; but in the exceptional cases, the class of solutions may be further restricted so as to obtain simple eigenvalues in the new space.)

Thus, the existence of nontrivial steady solutions for μ near μ_0 is proved; these solutions represent convection in the period cells. Further symmetry restrictions yield convection in hexagonal or triangular cells. It is believed that only the convection cells with $\ell_1 = \infty$ (rolls) are stable, as these are the ones seen experimentally.

A few variants on this problem should be mentioned. They all lead to similar results. One can envisage the upper plane being removed so that the upper surface is free. Then different boundary conditions should be imposed there; surface tension effects become important. This is the situation with Bénard's early experiment, (1900), in which hexagonal cells were seen. For an analytical treatment, see Pearson (1958).

The problem can be generalized by allowing more exact fluid dynamical equations rather than those of the Boussinesq approximation (Fife and Joseph (1969); Fife (1970/71)) and by allowing internal heating. Finally, substances may be dissolved in the fluid; then their diffusion and convection must be accounted for. This leads to a system like the Bossinesq system, but larger.

4. The Appearance of Periodic Solutions

Again, suppose the given family U^0 is time-independent. Then the linear operator A_μ is also time-independent. Suppose, for some $\mu_0 \in I$, that A_{μ_0} has a nonzero purely imaginary pair of simple complex-conjugate eigenvalues. Then Theorem 3 applies by virtue of Property 2 given in Section II.2, and we conclude that for μ in at least a one-sided neighborhood of μ_0, there exist periodic solutions provided that $\mathrm{Re}[B\phi] \neq 0$. For full details see Joseph and Sattinger (1972).

The verification of these assumptions in many particular cases would be difficult, yet in all likelihood, this type of bifurcation phenomenon is common. See Davey, DiPrima, and Stuart (1968) for a discussion of it and analysis of it in the case of a rotating cylinder problem. Bifurcation into periodic flows, and other aspects of bifurcation theory as well, is studied for plane Poiseuille flow in the papers of Chen and Joseph (1973), Jospeh and Chen (to appear), and Joseph (to appear).

5. Bifurcation into an Invariant Torus

The appearance of invariant tori for systems of ordinary differential equations has been well studied in the past by several people, possibly beginning with Bogoljubov and Krylov (1934). In particular, see Sacker (1965); also Hale (1969) for an excellent survey of known results.

For partial differential equations much less is known. One study is that of Iooss (to appear), which this section follows. We study the case when the given family U_μ^0 of solutions of (2.3), (2.2) are periodic in time with fixed period T. Then the operator B_μ in (2.5) will depend periodically on t. We operate within the space $\hat{B}$ defined on page 26 with $T_1 > T$. We examine the analytic solution operator $S_{\mu,T}$. It has a compact Frechet derivative T_μ. If the spectrum of T_μ is completely inside the unit circle, then the given periodic solution is stable and all small disturbances will die down in time.

However, suppose that as μ progresses through the value μ_0, certain eignevalues cross the unit circle. Then qualitatively different stable solutions may appear. Specifically, they may exist for μ in (at least) a one-sided neighborhood of μ_0. If a simple eigenvalue crosses at the point 1, then a new periodic solution will appear in addition to the old one. The same happens when complex pairs of eigenvalues cross where one of them is a root of unity; however in this case the new solutions may have a period which is a multiple of the old one.

Finally, consider when a pair of simple eigenvalues cross the unit circle at points ζ_0, $\overline{\zeta}_0$ not equal to a root of unity. Let us also assume there are no other eigenvalues on the circle besides these. Then under a certain circumstance, an invariant torus may occur. This condition has been elucidated by Iooss (to appear) on the basis of the results of Ruelle and Takens (1971); see also Lanford (1973). It takes the form: $\alpha \neq 0$, where α is a certain constant which can be expressed in terms of the eigenvalue ζ_0, its eigenvector, and the corresponding eigenvector of the adjoint operator. Again, these conditions would be difficult to verify in many actual cases, but they appear to be mild enough assumptions.

A word about the method of proof is in order. Let us denote by Φ_τ the map $S_{\mu_0+\tau,T}$, and by Ψ the map $(u_0,\tau) \to (\Phi_\tau u_0,\tau)$ defined for (u_0,τ) in a neighborhood of the origin in $\mathcal{D} \times \mathbb{R}$. Its derivative $D\Psi(0)$ has three eigenvalues: ζ_0, $\overline{\zeta}_0$, and 1, the latter arising from the second component of the map Ψ. The Center Manifold Theorem (for example, see Lanford (1973)) now states that there is a three-dimensional manifold M near the origin $\mathcal{D} \times \mathbb{R}$ which is locally invariant under Ψ. This means there is a neighborhood N_0 of the origin in $\mathcal{D} \times \mathbb{R}$ such that $(u_0,\tau) \in M$ and $\Psi(u_0,\tau) \in N_0 \Rightarrow \Psi(u_0,\tau) \in M$. It is immediate from the definition of Ψ that each section M_τ of M with $\tau = \text{const}$ is locally invariant for the mapping Φ_τ. For fixed τ, examine the mapping Φ_τ restricted to M_τ. Since M_τ is two-dimensional and invariant, we may introduce a coordinate system to obtain a mapping from a neighborhood of the origin in $\mathbb{R}^2$ into $\mathbb{R}^2$. Such mappings can be analyzed directly and a condition given under which there exists an invariant closed curve. This condition, translated back into the terms of the original problem, takes the form of $\alpha \neq 0$ and τ being in some one-sided neighborhood of the origin. We will thus have, for each such τ, an invariant closed curve Γ_τ on M_τ, hence in $\mathcal{D}$, for the map $\Phi_\tau = S_{\mu+\tau,T}$. This means that if the evolution problem (2.5) is solved with initial conditon $u = u_0 \in \Gamma_\tau$, then $u(nT) \in \Gamma_\tau$ for all positive integers n. The invariant two-dimensional torus will thus be made up of the trajectories of all solutions with initial values on Γ_τ. Since $\Gamma_\tau \to \{0\}$ as $\tau \to 0$, these tori shrink to a closed loop (representing the given periodic solution $U^0_{\tau_0}$) as $\tau \to \tau_0$.

The above analysis also yields results about the stability of the new solutions. The Center Manifold Theorem provides the fact that the invariant manifold M_τ is attracting, and the analysis of the two-dimensional mapping problem gives conditions under which Γ_τ is attracting. It is when $\alpha > 0$. Thus, if one is seeking stable invariant tori, the condition to be verified is $\alpha > 0$.

6. Branching in the Variational Theory of Turbulence

There is in existence another fluid dynamical model which, though simpler than the theory of the Navier-Stokes equations themselves, nevertheless exhibits a type of repeated supercritical branching into more and more complicated solutions reminiscent of the Landau conjecture. It is the variational theory of turbulence. (See Howard (1963), Busse (1969) and others). The branching aspects of this theory in the case of heat transport across a fluid-filled porous layer heated from below are analyzed by Joseph (1974); a similar analysis can be extended to the theory in other contexts. In Joseph's treatment, statistical stationary turbulent flow is assumed; that is, the horizontal average of the velocity is zero, and that of the temperature depends only on z (the vertical coordinate). Two parameters are important: the Rayleigh number R, proportional to the imposed temperature difference, and the discrepancy μ between the turbulent convective heat transport and that which would ensue from conduction alone.

The variational problem (or rather one of those considered) is to determine, for a given μ, the minimum value of R for which that μ is possible. Very roughly speaking, one assumes an expansion $\theta = \sum_{n=1}^{N} \theta_n(z) g_n(x,y)$ for the temperature fluctuation θ (and for other quantities as well), where the g_n are eigenfunctions of the Laplacian, $g_{xx} + g_{yy} + \alpha_n^2 g = 0$. There is a certain functional $\mathcal{F}_N(\theta_j, \alpha_j, \mu)$ of these α_j's and $\theta_j(z)$'s, $j \leq N$, which is to be minimized by choosing the θ_j's and α_j's appropriately. Let the minimum be $F_N(\mu)$. Then the desired minimum value of R is $F(\mu) = \operatorname*{Min}_{N \geq 1} F_N(\mu)$.

For small μ this minimum is achieved with $N = 1$. But for μ slightly larger than a critical number μ_0, it is attained with $N = 2$, so that θ is a combination of two of the θ_j's, with different wave numbers α_1, α_2. Presumably there will be a second critical number beyond which $N > 2$, and so on.

The complementary problem is, given R, what is the maximum μ that can be attained? Of course, the maximum μ is achieved with the same function θ as in the first problem. Thus, as R increases, the maximum heat exchange is brought about by more and more irregular functions $\theta(x,y,z)$, since they will be combinations of periodic functions of x and y with greater and greater numbers of incommensurate wave numbers involved.

P. C. Fife

III. PROBLEMS ARISING IN CONNECTION WITH CHEMICAL REACTION AND DIFFUSION

Mixtures of substances which diffuse and react are supposedly described by systems of coupled nonlinear diffusion equations. The object of this lecture is to touch upon certain topics in the branching theory of such systems. These equations are also relevant in a variety of areas besides those implied by the title. Examples are the generation of heat in a solid by periodic loading or ohmic heating by an electric current, and that in viscous fluid by internal friction.

Consider the system of n equations

$$\frac{\partial U}{\partial t} - D\Delta U = f(\mu,x,t,U), \tag{1.1}$$

where D is an $n \times n$ diagonal matrix (giving the diffusion coefficients) and μ is a real parameter. Assume some appropriate boundary conditions are prescribed on the boundary of a bounded region. Practically every general result and technique discussed in the lectures on fluid dynamics can be applied in this setting. One can study the bifurcation of families of steady solutions here just as was done in the other case, as well as the appearance of periodic solutions. Probably the existence of invariant tori could also be proved under suitable assumptions.

1. A Model Oscillatory Chemical Reaction

There has been considerable interest recently in chemical reactions which oscillate, and in reacting mixtures which display some sort of spacial structure. One possible approach to a study of such phenomena would be in terms of the branching phenomena we have outlined. Suppose, for example, that (1.1) has a given family of equilibrium states U^0_μ depending smoothly on μ. There might exist critical values of μ at which new steady states arise, possible with at least a semblance of spacial structure, or at which periodic solutions begin to appear. As an example, we consider the model reaction (Glansdorf and Prigogine (1971)).

$$A \to X$$

$$2X + Y \to 3X$$

$$B + X \to Y + D$$

$$X \to E$$

We suppose the concentration of the substances A, B, D, and E to be artificially held constant throughout the system, and the concentrations of X and Y to be fixed at α and μ/α, respectively, on the boundary, where α and μ are the imposed concentrations of A and B respectively. We consider the system to have one space dimension, with domain $0 \leq x \leq 1$. We also assume the products X and Y to diffuse. For simplicity, we shall take the diffusion coefficients and the reaction rates all to be equal to 1.

Letting U and V denote the concentrations of X and Y respectively, we have the system

$$U_t - U_{xx} - \alpha + (\mu + 1)U - U^2V = 0 \qquad (1.2a)$$

$$V_t - V_{xx} - \mu U + U^2V = 0. \qquad (1.2b)$$

We shall fix α and let μ be a perturbation parameter.

First we study the bifurcation of steady-state solutions. It may be checked that $U^0 \equiv \alpha$, $V^0 \equiv \mu/\alpha$ is a solution of (1.2). We call it the "uniform" solution. We seek other time-independent solutions of the form $U = U^0 + u$, $V = V^0 + v$. We then obtain the equations

$$u'' + (\mu - 1)u + \alpha^2 v + N_\mu^{(1)}(u,v) = 0,$$

$$v'' - \mu u - \alpha^2 v + N_\mu^{(2)}(u,v) = 0,$$

$$u = v = 0 \quad \text{for} \quad x = 0 \quad \text{or} \quad 1;$$

where the $N_\mu^{(i)}$ are quadratic terms. This system can be written in a form very close to (I.2.1):

$$(A + \mu B)\overline{U} + N_\mu(\overline{U}) = 0, \qquad (1.3)$$

where

$$A = \frac{d^2}{dx^2} + \begin{bmatrix} -1 & \alpha^2 \\ 0 & -\alpha^2 \end{bmatrix}, \qquad B = \begin{bmatrix} 1 & 0 \\ -1 & 0 \end{bmatrix},$$

and N is quadratic. In order to apply Theorem 2, we must find values of μ for which $A_\mu \equiv A + \mu B$ has a one-dimensional nullspace. Since this is an operator with constant coefficients, the nullvector $\Phi = (\phi, \psi)$ will be of the form

$$\phi = \phi_0 \sin n\pi x; \qquad \psi = \psi_0 \sin n\pi x.$$

Requiring this to be a nullvector of A_μ yields the relation

$$\mu = 1 + \alpha^2 + n^2\pi^2 + \frac{\alpha^2}{n^2\pi^2}, \tag{1.4}$$

which gives an infinite number $\mu^{(n)}$ of critical values of μ, corresponding to $n = 1, 2, 3, \ldots$. As the parameter μ is increased, we expect the uniform solution U^0, V^0 to lose stability at the smallest of the $\mu^{(n)}$, which will occur at some value of $n = n_c$. We assume there is a unique value n_c (true except for discrete values of α). The nullspace will then be one-dimensional. We denote by μ_0 the corresponding value of μ determined from (1.4).

According to Theorem 2, the only thing left to check to determine whether a bifurcation actually occurs, is

$$PB\Phi \neq 0, \tag{1.5}$$

where P is a projection onto the complement of the range of A_{μ_0}. To calculate this projection, we must know the nullvector of the adjoint operator. Accordingly, denote by Φ^* the nullvector of $A^* + \mu_0 B^*$.

Again, since the adjoint has constant coefficients, Φ^* will be the form

$$\Phi^* = (\phi_0^* \sin n_c \pi x, \psi_0^* \sin n_c \pi x).$$

Let X be the two-vector $\begin{bmatrix}\phi_0\\ \psi_0\end{bmatrix}$, and X^* the vector $\begin{bmatrix}\phi_0^*\\ \psi_0^*\end{bmatrix}$. Then the condition (1.5) may be written in the form $BX \cdot X^* \neq 0$, in the sense of ordinary scalar products. But a glance at the matrix B shows that this may in turn be written $\phi_0(\phi_0^* - \psi_0^*) \neq 0$, and a routine calculation shows that $\phi_0 \neq 0$ and $\phi_0^* - \psi_0^* \neq 0$.

Thus the condition is indeed fulfilled, and the existence of non-trivial solutions for μ near μ_0 is guaranteed. These solutions will be approximately sinusoidal, with $n_c - 1$ zeros. Nicolis and Auchmuty (1974) and Boa (1974) have made a stability analysis of this bifurcation, and Boa has analyzed the time evolution to the new steady state, as well as an analysis of the onset of periodic motion, as described below. These authors do not restrict the diffusion coefficients to be one.

We now ask whether periodic solutions can arise as bifurcation branches from the uniform steady solution. According to Theorem 3, this will happen if stability is lost by means of a pair of simple eigenvalues of the operator A_μ crossing the imaginary axis. If this happens, then for some μ and some $\omega > 0$, the spectrum of A_μ will consist of discrete eigenvalues with negative real part, except for a pair $\pm i\omega$ of simple ones. Suppose this is true, and let $\Phi = (\phi, \psi)$ be the nullvector corresponding to $i\omega$. As before, it will be of the form $\phi = \phi_0 \sin n\pi x$, $\psi = \psi_0 \sin n\pi x$, where now ϕ_0, ψ_0 will be complex numbers.

The condition $(A_\mu - i\omega I)\Phi = 0$ reduces to

$$\begin{bmatrix} -(n^2\pi^2 + i\omega + 1) & \alpha^2 \\ 0 & -(n^2\pi^2 + i\omega + \alpha^2) \end{bmatrix} X + \begin{bmatrix} 1 & 0 \\ -1 & 0 \end{bmatrix} X = 0.$$

Setting the determinant equal to zero and solving for μ, we have

$$\mu = 1 + \alpha^2 + n^2\pi^2 + i\omega + \frac{\alpha^2}{n^2\pi^2+i\omega} . \tag{1.6}$$

Since μ is real, we have

$$0 = \mathrm{Im}\mu = \omega\frac{n^4\pi^4+\omega^2-\alpha^2}{n^4\pi^4+\omega^2},$$

which implies that either

(a) $\omega = 0$, or

(b) $\omega^2 = \alpha^2 - n^4\pi^4$.

In case (a), we are back to the case of steady state bifurcation, treated above, for which the critical value μ_0 has already been determined.

In case (b), which can only happen if $\alpha \geq n^2\pi^2$, we have from (1.6)

$$\mu = \hat{\mu}^{(n)} = \mathrm{Re}\hat{\mu}^{(n)} = 1 + \alpha^2 + n^2\pi^2 + \frac{\alpha^2 n^2\pi^2}{n^4\pi^4+\omega^2} = 1 + \alpha^2 + 2n^2\pi^2.$$

Clearly the minimal value $\hat{\mu}_0$ of $\hat{\mu}^{(n)}$ is that corresponding to $n = 1$:

$$\hat{\mu}_0 = 1 + \alpha^2 + 2\pi^2.$$

If $\alpha \leq \pi^2$, then only case (a) can occur, and we conclude that stability is lost to a new steady state solution with a certain number $n_c - 1$ of zeros. We say in this case that the "principle of exchange of stabilities" holds.

On the other hand if $\alpha > \pi^2$, then periodic solutions may arise, and in fact will, if $\hat{\mu}_0 < \mu_0$. But this inequality is guaranteed to be true, because

$$\mu_0 = \min_{n=\text{integer}} \left(1 + \alpha^2 + n^2\pi^2 + \frac{\alpha^2}{n^2\pi^2}\right) \geq 1 + \alpha^2 + \min_{\theta \text{ real, positive}} \left(\theta + \frac{\alpha^2}{\theta}\right)$$

$$= 1 + \alpha^2 + 2\alpha > 1 + \alpha^2 + 2\pi^2 = \hat{\mu}_0.$$

Thus the principle of exchange of stabilities does not hold, and a periodic motion arises, with wave form like $\sin \pi x$, and frequency $\omega = (\alpha^2 - \pi^4)^{1/2}$.

2. Travelling Wave Solutions

Another interesting phenomenon which has been observed in certain chemical mixtures is that of periodic travelling waves. Kopell and Howard (1973) have made an investigation of the mathematics behind this. We present here some aspects of their work. (See also Howard and Kopell (to appear)). Mathematically speaking, such waves would be represented by solutions of (1.1) of the form $U = u(ct - \alpha \cdot x)$ for some unit vector α. Here c is the speed of the wave. For purposes of illustration we take the simplest case, namely the case when D is a scalar matrix: $D = kI$. For convenience we also restrict to one space dimension. We further assume that f is independent of x, t, and μ, and that $f(0) = 0$. The concept of branching of solutions, of course, depends on the existence of a variable parameter such as μ, so it may appear strange to disregard such parameters as we propose to do here. Nevertheless in the present circumstance, the need for a parameter is fulfilled by the wave speed c.

The condition $f(0) = 0$ means that the problem always has a zero solution ($U = 0$ is an equilibrium state), and we are seeking a bifurcation into the realm of nontrivial travelling waves. Of course, periodic waves near other equillibrium states could be studied as easily.

Introducing the indicated simplifications into (1.1) and writing $U = u(\tau)$, with $\tau = (ct - x)$, we have

$$cu' - ku'' = f(u). \tag{2.1}$$

By defining $v = u'$, $w = \binom{u}{v}$ (a vector with $2n$ components), the system (2.1) could be put into the form

$$\frac{dw}{d\tau} + (A + cB)w + Nw = 0, \tag{2.2}$$

which is the type treated earlier (I.3.1), with A and B certain matrices. Then the analysis (according to Theorem 3) would proceed by finding a value c_0 of c for which $A + c_0B$ has a pair of complex conjugate simple imaginary eigenvalues $\lambda = \pm i\omega_0$.

If c_0 is such a value, then the linear operator $A + cB$ would be such that $\lambda I + (A + c_0B)$ is singular. In simple terms, the differentiation operator $\frac{d}{d\tau}$ is replaced by λ, the problem is linearized, and the resulting operator is required to be singular. It proves to be more convenient to do exactly this from the start, namely in the context of (2.1) rather than (2.2). So doing, we obtain

$$c_0\lambda I - k\lambda^2 I - M \text{ is singular,} \tag{2.3}$$

where the matrix $M = f'(0)$.

At this point we assume that M has a pair of eigenvalues of the form $p \pm iq$, with $p > 0$, $q > 0$, and no others with real part p. Then the matrix $M - \alpha^2 kI$ has eigenvalues $p - \alpha^2 k \pm iq$. We use this result to solve (2.3) by setting the eigenvalue with the "+" sign equal to $c_0\lambda$:

$$c_0\lambda = p - \alpha^2 k + iq, \tag{2.4}$$

and also setting $\alpha^2 = -\lambda^2$.

Since we are looking for imaginary eigenvalues $\alpha^2 = -\omega_0^2$, we obtain $\alpha^2 = \omega_0^2$, and $ic_0\omega_0 = p - \omega_0^2 k + iq$. From this, we obtain

$$\omega_0 = q/c_0 \quad \text{and} \quad p - (\frac{q^2}{c_0^2})k = 0,$$

hence $c_0^2 = q^2k/p$.

We have found that for the indicated value of $c = c_0$, the operator $A + c_0B$ has eigenvalues $\pm i\omega_0 = \pm iq/c_0$. From Theorem 3, we conclude that (2.2), hence (2.1), has nontrivial periodic solutions with frequencies near ω_0 for certain values of c near c_0, (typically, those in a sufficiently small one-sided neighborhood of c_0), provided that $\mathrm{Re}\lambda'(c) \neq 0$ at $= c_0$.

To check this last condition, we go back to (2.4) and write

$$c\lambda = p + \lambda^2k + iq,$$

hence

$$\lambda = \frac{c-(c^2-4k(p+iq))^{1/2}}{2k}.$$

From this it may be verified that $\mathrm{Re}\frac{d\lambda}{dc} \neq 0$ at $c = c_0$.

In Kopell and Howard (1973), the case when D is not a scalar matrix (in fact, not even diagonal; only positive definite) is treated. It was found, roughly speaking, that there exists a critical value c_0 from which periodic solutions bifurcate, provided that D is sufficiently near to being a scalar matrix. They also prove the existence of plane waves with large wave-length near limit cycles of the corresponding kinetic equations (with D set equal to 0). The same authors (see

Howard and Kopell, (to appear)) also make interesting studies of the interaction of plane waves; in particular, they treat the case of two different wave trains separated by a "shock".

3. Transition Layers

The following is a simple prototype of a class of perturbation problems arising in several applications:

Find a scalar function $u(x,\varepsilon)$, defined for $-\infty < x < \infty$ and $0 < \varepsilon < \varepsilon_0$, satisfying the equation

$$u_{xx} = f(u, \varepsilon x, \varepsilon) \tag{3.1}$$

and the boundary conditions

$$u(-\infty,\varepsilon) = 0, \qquad u(\infty,\varepsilon) = 1. \tag{3.2}$$

We assume f is a smooth function of all its arguments.

Infinite interval two-point boundary value problems with a small parameter, such as this, occur in the study of transition layers in singular perturbation problems (which could be interpreted as a type of internal structure in steady-state systems of the type (1.1)), in the study of shocks, and in the theory of travelling plane wave solutions of (1.1) (not necessarily periodic). Such waves could represent, for example, combustion fronts, oscillating chemical reactions, nerve impluse propagation, or propagation of traits or fads in a population. Not all of these problems may be solved by the method presented. Lack of time prohibits discussion of these applications; however we shall comment on the first of them later.

A naive first approximation to (3.1) is obtained by setting $\varepsilon = 0$, $u(x,0) = u^0(x)$:

$$u^0_{xx} = f(u^0,0,0); \qquad u^0(-\infty) = 0, \qquad u^0(\infty) = 1. \tag{3.3}$$

Since a solution satisfying these boundary conditions necessarily satisfies $u^0_x = u^0_{xx} = 0$ at $x = \pm\infty$, we require

$$f(0,0,0) = f(1,0,0) = 0. \tag{3.4}$$

Furthermore the conditions

$$f_u(0,0,0) > 0, \qquad f_u(1,0,0) > 0 \tag{3.5}$$

guarantee u^0's approach to its limits to be exponential. Now multiplying (3.3) by u^0_x and integrating with respect to x from $-\infty$ to ∞ yields the following necessary and sufficient condition for the existence of a solution:

$$\int_0^1 f(u,0,0)du = 0; \qquad \int_0^k f(u,0,0)du > 0, \qquad k \in (0,1). \tag{3.6}$$

We note the following feature of the problem (3.3): Under the conditions for existence that we have given there exist an infinite number of solutions u^0, for the independent variable x may be translated at will. This feature is especially noteworthy since it contrasts with the fact (as we shall see) that (3.1), (3.2) will in general (if f really depends on its second argument) have a unique solution for $\varepsilon \neq 0$. This situation is somewhat reminiscent of the example in Section I.4, in which the base problem also has an infinite number of solutions, and the branch of solutions of the nonreduced problem, parameterized by ε, emanates (as it does in this case) from a single one of them. Part of the problem is to determine which one of the base solutions serves as "branch point". The difference is that in the present circumstance, the base problem is nonlinear; but on the other hand, its set of solutions may be characterized very simply as the set $\{u^0(x - \mu):\mu \in \mathbb{R}\}$, where $u_0(x)$ is the solution (say) satisfying $u^0(0) = \frac{1}{2}$. It should be remarked that this is not a bifurcation problem in the strictest sense of that term, but it does have important features in common with such problems.

We proceed as follows to determine an exact solution of (3.1), (3.2). We change variables, setting $\xi = x - \mu$, where $\mu(\varepsilon)$ is a function to be determined. We also define $\lambda(\varepsilon) = \varepsilon\mu(\varepsilon)$, and seek a solution in the form

$$u(x,\varepsilon) = u^0(\xi) + v(\xi,\varepsilon),$$

where $v \to 0$ as $\varepsilon \to 0$. The principal term $u^0(\xi) = u^0(x - \mu)$ will then be "centered" at $x = \mu(\varepsilon)$, since that is where u^0 is the average $(\frac{1}{2})$ of its limiting values at $\pm\infty$. To keep this interpretation of μ and to have $u(x,\varepsilon)$ centered at $x = \mu$, we would logically want to have $v = 0$ at $\xi = x - \mu = 0$. This could be a sort of normalization condition on v, to go with the interpretation of μ. However, the mathematics turns out to be easier if we impose an alternate normalization condition on v, to be given below.

Rewriting (3.1) in terms of ξ and v, we obtain

$$v_{\xi\xi} = [f(u^0(\xi) + \varepsilon v, \varepsilon\xi + \lambda, \varepsilon) - f(u^0(\xi),0,0)] \equiv h(v,\xi,\lambda,\varepsilon), \tag{3.7}$$

$$v(\pm\infty, \varepsilon) = 0.$$

Define the operator $F(v,\lambda,\varepsilon) \equiv v_{\xi\xi} - h(v,\xi,\lambda,\varepsilon)$ and write (3.7) as

$$F(v,\lambda,\varepsilon) = 0. \tag{3.8}$$

It is verified by differentiating (3.3) that the function $\phi(\xi) \equiv u^{0'}(\xi)$ satisfies

$$F_u(0,0,0;\phi) \equiv \phi'' - h_v(0,\xi,0,0)\phi = 0, \qquad \phi(\pm\infty) = 0.$$

Thus ϕ is a nullvector of $F_u(0,0,0)$. Since it is positive, we know from the spectral theory of ordinary differential equations that 0 is a simple eigenvalue and the lowest point on the spectrum of $-F_u(0,0,0)$; also since $h_v(0,\pm\infty,0,0) > 0$ by (3.5), we have that the rest of the spectrum is bounded away from 0 (see Fife (1974)).

Using this, we can show that our basic Theorem 1 (page 5) is applicable to the problem (3.8). Specifically, we let B be the Banach space $C^2(-\infty,\infty) \cap H_2(-\infty,\infty)$; $B_0 = C^0(-\infty,\infty) \cap H_0(-\infty,\infty)$, and consider F as a mapping: $B \to B_0$. By virtue of the above facts about the spectrum, we know that $F_u(0,0,0)$ has nullspace spanned by ϕ, and range characterized by $\{f \in B_0 : (f,\phi) = 0\}$, where (f,ϕ) is the scalar product in $L_2(-\infty,\infty)$. We denote the range by Z, $X = \{v \in B : (v,\phi) = 0\}$, and $Y = \text{span}\ \{\phi\}$. Then $W = B_0 = Y \oplus Z$.

As normalization condition referred to above, we require $(v,\phi) = 0$, so that $v \in X$. Finally, we set $\Lambda = \mathbb{R}$.

Thus F maps $X \times \Lambda \times \mathbb{R}$ into $W = Y \oplus Z$, and $F_u(0,0,0)$ is an invertible map from X onto Z, so assumption (ii) of Theorem 1 is satisfied. Clearly, so are (i) and (iv).

To check (iii), we calculate

$$F_\lambda(0,0,0;\lambda) = -h_\lambda(0,\xi,0,0)\lambda = f_2(u^0(\xi),0,0)\lambda,$$

so that $PF_\lambda\lambda = \lambda(\phi,F_\lambda(0,0,0))\phi$. This is an invertible map from $\Lambda = \mathbb{R}$ onto Y if and only if $(\phi,F_\lambda(\ldots)) \neq 0$; that is,

$$0 \neq \int_{-\infty}^{\infty} \phi(\xi)F_\lambda(\ldots)d\xi = \int_{-\infty}^{\infty} u^{0\prime}(\xi)f_2(u^0(\xi),0,0)d\xi = \int_0^1 f_2(u,0,0)du. \tag{3.9}$$

Thus, we have:

THEOREM. _Under assumptions_ (3.4), (3.5), (3.6), _and_ (3.9), _there exist_ (_for small_ ε) _continuous functions_ $\mu(\varepsilon)$ and $u(x,\varepsilon) = u^0(\xi)$

$+ v(\xi,\varepsilon)$, <u>where</u> $\xi = x - \mu(\varepsilon)$, <u>satisfying</u> (3.1), (3.2), $v(\xi,0) = 0$. <u>Moreover</u> $(v,u_0') = 0$ <u>for all</u> ε.

As an immediate application, consider the singular perturbation problem

$$\varepsilon^2 u_{yy} = f(u,y,\varepsilon). \tag{3.9}$$

If f satisfies (3.4), then two obvious solutions of the reduced problem (with $\varepsilon = 0$) are $u \equiv 0$ and $u \equiv 1$. We also assume (3.5). One may ask the question, does there exist an exact solution exhibiting a sharp transition between these two reduced solutions? To answer this, we construct a function $J(y) = \int_0^1 f(u,y,0)du$. The answer is yes, if there exists a point y_0 such that $J(y_0) = 0$, $J'(y_0) \neq 0$. For with no loss, we may suppose $y_0 = 0$, define the stretched variable $x = y/\varepsilon$, and the problem becomes (3.1), (3.2), with all conditions for existence satisfied. The transition will be in a small interval on the y-axis with width of order ε.

This type of problem was solved by other methods by Vasil'eva and Butuzov (1973). For a generalization to other types of ordinary differential equations, and for other information, see Fife (1974). Equation (3.9) could represent a type of steady-state model for a reacting and diffusing system, with small diffusion coefficient. As such, the existence of such transition layers would represent a type of internal structure to such a system of a kind quite different from that studied in

Section III.1. It would be expected that systems of equations of the type (3.9), with small and large parameters of differing orders of magnitude, may be capable of exhibiting more complicated kinds of internal structure; these lines could even prove fruitful to study in connection with the appearance of spacial structure in biological organisms. In the next section we shall indicate an analogous result in two space dimensions.

4. The Plateau Phenomenon

This will be a brief discussion, following Fife and Greenlee (1974), of a generalization to a partial differential equation of the problem considered in the foregoing section. Consider the problem

$$\varepsilon^2 \Delta u = f(u,x,\varepsilon), \quad x \in \Omega \text{ (bounded)} \subset \mathbb{R}^2, \tag{4.1}$$

$$u|_{\partial\Omega} = h(x). \tag{4.2}$$

Suppose the reduced problem $f(u,x,0) = 0$ has two distinct solutions $u = g_1(x)$, $u = g_2(x)$, with $f_u(g_i(x),x,0) > 0$. As before, we define

$$J(x) = \int_{g_1(x)}^{g_2(x)} f(u,x,0)du.$$

Suppose there exists a closed curve $\Gamma \subset \Omega$ on which $J(x) = 0$, and which separates Ω into two subdomains Ω_1 and Ω_2 as shown.

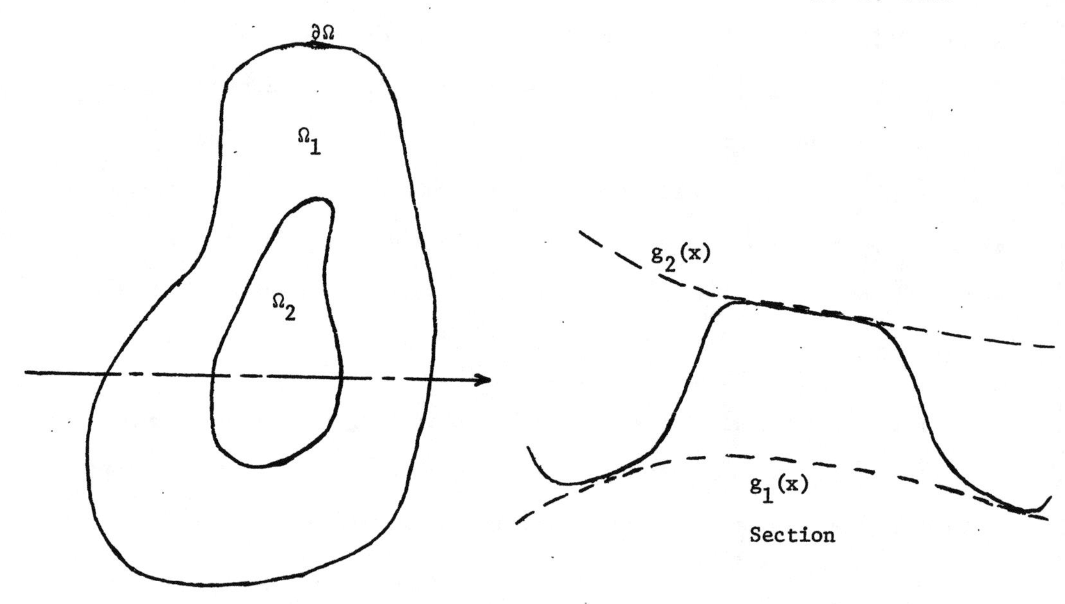

We also assume $\int_{g_1(x)}^{k} f(u,x,0)du > 0$ for $k \in (g_1(x),g_2(x))$, $x \in \Gamma$, and that $\partial_n J(x) < 0$ on Γ, this denoting the derivative normal to Γ from Ω_1 into Ω_2.

The above bifurcation method (much complicated in this case) can be applied to prove the existence of a solution approaching $g_1(x)$ in Ω_1 and $g_2(x)$ in Ω_2 (see the section drawing above).

The complications involve the following, among many others: λ now is not just a scalar, but a function $\lambda(s,\varepsilon)$, where s is arclength along the curve Γ. Thus the parameter space Λ in Theorem 1 will have as elements periodic functions of a single variable s. Furthermore, Theorem 1 as it stands is not applicable. It must be modified so that the space X, Y, Z, and Λ will themselves depend on ε. Thus for each ε, $F(\cdot,\cdot,\varepsilon)$ will be a mapping from $X_\varepsilon \times \Lambda_\varepsilon$ into W_ε. The inverses of the invertible operators mentioned in Assumptions (ii) and (iii) may be unbounded in ε as $\varepsilon \to 0$, as long as the behavior is not too singular.

These changes require a nontrivial modification of the implicit function theorem. Moreover, the verification of the assumptions of this modified theorem is very difficult in the present case, and requires a very careful choice of the spaces X_ε, Λ_ε, etc.

P. C. Fife

REFERENCES

H. Bénard, Rev. Générale Sci Pures Appl 11 (1900).

M. S. Berger and P. C. Fife. Bull. Amer. Math. Soc., 72 (1966).

M. S. Berger, Comm. Pure Appl. Math. 21 (1968)

Comm. Pure Appl. Math. 20 (1967).

J. A. Boa, Ph. D. Thesis, Calif. Inst. Tech. (1974). See also forthcoming paper by J. A. Boa and D. S. Cohen.

N. Bogoljubov and N. Krylov, C. R. Acad. Sci. Paris 199 (1934), 1592-3.

F. Busse, J. Fluid Mech. 37 (1969).

T. S. Chen and D. D. Joseph, J. Fluid Mech. 58 (1973).

M. G. Crandall and P. H. Rabinowitz, J. Functional Anal. 8 (1971).

M. G. Crandall and P. H. Rabinowitz, Ibid. (to appear).

A. Davey, R. C. DiPrima and J. T. Stuart, J. Fluid Mech. 31 (1968).

P. Fife and D. Joseph, Arch. Rat. Mech. Anal. 33 (1969).

P. Fife, Indiana Univ. Math. J. 20 (1970/71).

J. Differential Equations 15 (1974).

P. Fife and W. M. Greenlee, Usp. Matem. Nauk SSSR, (1974) (to appear).

G. Foias and G. Prodi, Rend. Sem. Mat. Univ. Padova 39 (1967).

P. Glansdorff and I. Prigogine, Thermodynamics of Structure, Stability and Fluctuations, Wiley, New York (1971).

J. K. Hale, Ordinary Differential Equations, Wiley, New York (1969).

E. Hopf, Berichten der Math-Phys. Klasse der Sächischen Akad. Wiss. Leipzig, XCIV (1942).

Comm. Pure Appl. Math. 1 (1948).

L. N. Howard, J. Fluid Mech. 17 (1963).

L. N. Howard and N. Kopell, Article in Proceedings of Symposia in Applied Mathematics, Vol. 27, Amer. Math. Soc. (1974) (to appear).

G. Iooss, C. R. Acad. Sci. Paris 273, Series A (1971).

G. Iooss, Arch. Rat. Mech. Anal. 47 (1972).

Bifurcation et Stabilité, Lecture Notes No. 31, Université Paris XI (1972-73).

Arch. Rat. Mech. Anal. (to appear).

D. D. Joseph, J. Fluid Mech. 47 (1971).

D. D. Joseph and D. H. Sattinger, Arch. Rat. Mech. Anal. 45 (1972).

D. D. Joseph, Article in Nonlinear Problems in the Physical Sciences and Biology, Lecture Notes in Mathematics, Vol. 322, Springer, New York (1973).

Arch. Rat. Mech. Anal. 53 (1974).

Advances in Applied Mathematics, 14 (to appear).

D. D. Joseph and T. S. Chen, J. Fluid Mech. (to appear)

V. I. Judovič, Prikl. Mat. Meh. 30 (1966)=PMM 30 (1966).

Dokl. Akad. Nauk SSSR 169 (1966)

Prikl. Mat. Meh. 31 (1967).

Dokl. Akad. Nauk SSR 195 (1970)

Prikl. Mat. Meh. 35 (1971).

K. Kirchgässner, Z. Angew. Math. Phys. 12 (1961).

K. Kirchgässner and P. Sorger, Proc. Twelfth Inter. Congress.Appl. Mech, Stanford (1968).

Quart. J. Mech. Appl. Math. 32 (1969).

K. Kirchgässner and H. Kielhofer, Rocky Mtn. Jour. of Math. 3 (1973).

N. Kopell and L. N. Howard, Studies in Appl. Math. 52 (1973).

M. Krasnosel'skii, Topological Methods in the Theory of Nonlinear Integral Equations. Pergamon, New York (1964).

O. A. Ladyženskaja, The Mathematical Theory of Viscous Incompressible Flow. Gordon and Breach, New York (1963).

Dokl. Akad. Nauk SSSR 205 (1972).

L. Landau, C. R. Acad. Sci. USSR, 44 (1944).

L. Landau and E. Lifschitz, Fluid Mechanics, Oxford: Pergamon (1959).

O. Lanford, III, Article on Nonlinear Problems in the Physical Sciences and Biology. Lecture Notes in Mathematics, Vol. 322, Springer, New York (1973).

G. Nicolis, Article in Nonlinear Problems in the Physical Sciences and Biology. Lecture Notes in Mathematics, Vol. 322, Springer, New York, (1973).

Article in Proceeding of Symposia in Applied Mathematics, Vol. 27, Amer. Math. Soc. (1974).

G. Nicolis and G. Auchmuty, Proc. Nat. Acad. Sci., (U.S.A.) 71 (1974).

J. R. A. Pearson, J. Fluid Mech. 4 (1958).

G. H. Pimbley, Jr. Eigenfunction Branches of Nonlinear Operators and their Bifurcations, Lecture Notes in Mathematics, Vol. 104, Springer, New York (1969).

P. Rabinowitz, Arch. Rat. Mech. Anal. 29 (1968).

Rocky Mountain Jour. of Math. 3 (1973).

D. Ruelle and F. Takens, Comm. Math. Phys. 20, 167-192 (1971)

R. J. Sacker, Comm. Pure Appl. Math. 18 (1965).

D. Sather, Rocky Mountain Jour. of Math. 3 (1973).

D. H. Sattinger, Arch. Rat. Mech. Anal. 41 (1971).

Topics in Stability and Bifurcation Theory, Lecture Notes in Mathematics, Vol. 309, Springer, New York (1973).

Article in Nonlinear Problems in the Physical Sciences and Biology, Lecture Notes in Mathematics, Vol. 322, Springer, New York, (1973).

I. Stakgold, SIAM Review, 13 (1971).

G. I. Taylor, Philos. Trans. Roy. Soc. London Ser. A223 (1923).

M. M. Vainberg and V. A. Trenogin, Uspehi Matem, Nauk SSSR, 17 (1962).

M. M. Vainberg and V. A. Trenogin, Theory of Branching of Solutions of Nonlinear Equations (Russian), Nauka, Moscow, (1969).

A. B. Vasil'eva and V. F. Butuzov, Asymptotic Expansions of Solutions of Singularly Perturbed Equations, (Russian), Nauka, Moscow (1973).

O. Vejvoda, Czech. Math. Jour. 14 (89) (1964)

M. G. Velarde, Article in Hydrodynamics, Gordon and Breach, New York, (1974) (to appear).

W. Velte, Arch. Rat. Mech. Anal. 16 (1964).

Arch. Rat. Mech. Anal. 22 (1966).

CENTRO INTERNAZIONALE MATEMATICO ESTIVO

(C. I. M. E.)

THE THEORY OF CLOSED GEODESICS

W. KLINGENBERG

Corso tenuto a Varenna dal 16 al 25 giugno 1974

The Theory of closed geodesics

Six lectures by W. Klingenberg given at the C.I.M.E Advanced Study Institute on "Eigenvalues in nonlinear problems" in Varenna, June 16 - June 25, 1974.

Survey articles covering a major part of these lectures are:

[1] S.I. Alber, The topology of functional manifolds and the calculus of variations in the large, Uspekhi Mat. NAUK 25, 4 (1970), 57-122 (Russian).- Russ. Math. Surveys 25, 4 (1970), 51-117.

[2] W. Klingenberg, Closed geodesics, Berichte aus dem Mathematischen Forschungsinstitut Oberwolfach, Heft 4, 77-103, Bibliographisches Institut, Mannheim-Wien-Zürich, 1971.

[3] W. Klingenberg, Closed geodesics on riemannian manifolds, Proc. 13th Biennial Sem. Can. Math. Congress, pp. 69-92. Can. Math. Congress, Montreal, Can., 1972.

1. The Hilbert manifold ΛM of closed curves on a riemannian manifold M.

We consider a compact riemannian manifold M. The scalar product on M shall be denoted by $\langle\ ,\ \rangle$; we suppose that it is of class C^∞. The covariant derivative derived from $\langle\ ,\ \rangle$ will be denoted by ∇. By $S = [0,1]/\{0,1\}$ we denote the parametrized circle

1.1 DEFINITION.

$$C^0(S,M) := \{c : S \to M; \quad c \in C^0\}$$

$$C^\infty(S,M) := \{c : S \to M; \quad c \in C^\infty\}$$

$$H^1(S,M) := \{c : S \to M; \quad c \in H^1\}$$

here, H^1 is the class of absolutely continuous maps (i.e., the derivative exists almost everywhere) with square integrable derivative.

Note: $$C^\infty \subset H^1 \subset C^0$$

In C^0 we have naturally the compact-open topology. We will show that $H^1(S,M)$, which we will denote also by ΛM, carries canonically the structure of a Hilbert manifold. To see this we consider for $c \in C^\infty(S,M)$ the bundle induced from the tangent bundle $\tau : TM \to M$ of M :

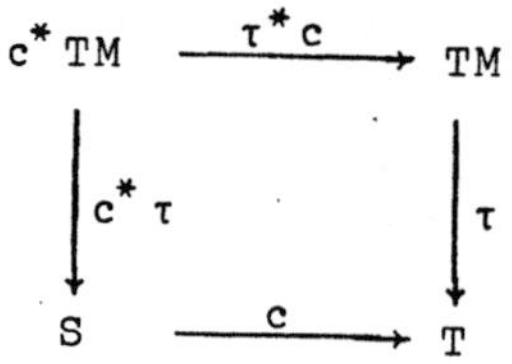

$$\begin{array}{ccc} c^*TM & \xrightarrow{\tau^* c} & TM \\ \downarrow{\scriptstyle c^*\tau} & & \downarrow{\scriptstyle \tau} \\ S & \xrightarrow{c} & T \end{array}$$

We put

$$H^0(c^*TM) := \{\xi : H^0\text{-section in } c^*\tau\}$$
$$H^1(c^*TM) := \{\xi : H^1\text{-section in } c^*\tau\}$$

We define a Hilbert space structure on these sets by taking as scalar product:

$$<\xi,\eta>_0 := \int_S <\xi(t),\eta(t)>\, dt \qquad \text{and}$$

$$<\xi,\eta>_1 := <\xi,\eta>_0 + <\nabla\xi,\nabla\eta>_0$$

respectively.

On

$$C^0(c^*TM) := \{\xi\colon C^0\text{-section in } c^*\tau\}$$

we define a Norm by

$$\|\xi\|_\infty := \sup_{t\in S}|\xi(t)|$$

The derived norms on $H^0(c^*TM)$ and $H^1(c^*TM)$ will be denoted by $\|\xi\|_0$ and $\|\xi\|_1$, respectively.

1.2 PROPOSITION. _The inclusions_

$$H^1(c^*TM) \hookrightarrow C^0(c^*TM) \hookrightarrow H^0(c^*TM)$$

are continuous. _More precisely_:

(i) _For_ $\xi \in C^0 : \|\xi\|_0 \leq \|\xi\|_\infty$

(ii) _For_ $\xi \in H^1 : \|\xi\|_\infty^2 \leq 2\|\xi\|_1$

We recall the following result from riemannian geometry, concerning the exponential map $\exp : TM \to M$.

1.3 LEMMA. <u>There exists an open neighborhood</u> $\mathcal{O}$ <u>of the</u> 0 -<u>section of</u> $\tau : TM \to M$ <u>such that the map</u>

$$(\tau, \exp) : \mathcal{O} \longrightarrow M \times M$$

<u>is a diffeomorphism onto a neighborhood of the diagonal of</u> $M \times M$. <u>In particular</u>, $\exp | \mathcal{O} \cap T_pM$ <u>is injective</u>.

Let now $c \in C^\infty(S,M)$. With $\mathcal{O}$ as in (1.3) we denote $c^*\mathcal{O} \subset c^*TM$ by $\mathcal{O}_c$. We write $H^1(\mathcal{O}_c)$ for the $\xi \in H^1(c^*TM)$ with $\xi(t) \in \mathcal{O}_c$.

1.4 DEFINITION.

$$\exp_c : H^1(\mathcal{O}_c) \to H^1(S,M) = \Lambda M$$
$$\xi = (\xi(t)) \to \exp \circ \xi(t)$$

1.5 PROPOSITION. $H^1(\mathcal{O}_c)$ <u>is an open set of the Hilbert space</u> $H^1(c^*TM)$ <u>and the map</u> (1.4) <u>is injective</u>.

We will use the maps (1.4) as charts for ΛM: We put $\exp_c H^1(\mathcal{O}_c) = \mathcal{U}(c)$. Clearly, the $\mathcal{U}(c)$, $c \in C^\infty(S,M)$, form a covering of ΛM . The next proposition shows that the

$(\exp_c^{-1}, \mathcal{U}(c)), c \in C^\infty(S,M)$, actually form a differentiable atlas of class C^∞ .

1.6 PROPOSITION. _Let_ $c,d \in C^\infty(S,M)$. _Then_

$$\exp_d^{-1} \circ \exp_c : \exp_c^{-1}(\mathcal{U}(c) \cap \mathcal{U}(d)) \longrightarrow \exp_d^{-1}(\mathcal{U}(d) \cap \mathcal{U}(c))$$

is a diffeomorphism between open sets of the separable infinite dimensional Hilbert spaces.

The proof uses in an essential way a result of Palais. Combining these results we have the

1.7 THEOREM. $\Lambda M = H^1(S,M)$ _is a Hilbert manifold with a canonical atlas given by_

$$(\exp_c^{-1}, \mathcal{U}(c)) \ , \quad c \in C^\infty(S,M) \ .$$

To show that we have on ΛM a canonical riemannian (or hilbertian) metric we first introduce the following two Hilbert bundles over ΛM:

$$\alpha^r : H^r(S,TM) \longrightarrow H^1(S,M) \ , \quad r = 0,1$$
$$(X(t)) \longmapsto (\tau X(t))$$

The bundle structure of α^0, α^1 is given naturally, using the canonical atlas of ΛM:

$$H^1(\mathcal{O}_c) \times H^r(c^*TM) \longrightarrow (\alpha^r)^{-1}\mathcal{U}(c)$$

is given by

$$(\xi(t),\eta(t)) \longrightarrow D_2 \exp(\xi(t))\cdot\eta(t)$$

where $D_2 := D \mid \text{fibre}$.

α^1 is canonically isomorphic to the tangent bundle $\tau_{\Lambda M}$ of $\Lambda M = H^1(S,M)$.

With this we have the

1.8 LEMMA.

(i)
$$\partial : H^1(S,M) \longrightarrow H^0(S,TM)$$
$$e = (e(t)) \longmapsto \partial e = (\dot{e}(t))$$

<u>is a</u> C^∞ <u>section in</u> α^0 .

(ii) <u>The Hilbert bundle</u> α^0 <u>has a riemannian metric</u> $\langle\, , \,\rangle_0$; <u>for</u> $c \in C^\infty(S,M)$ <u>is this given by the scalar product on</u> $H^0(c^*TM)$.

Using this we get:

1.9 THEOREM. _The Hilbert bundle_ $\alpha^1 = \tau_{\Lambda M}$ _has a riemannian metric_ $\langle\ ,\ \rangle_1$; _for_ $c \in C^\infty(S,M)$ _it is given by the scalar product on_ $H^1(c^*TM)$.

A further structure on ΛM is given by the energy integral

$$E : \Lambda M \longrightarrow T : c \longmapsto \frac{1}{2} \langle \partial c, \partial c\rangle_0$$

1.10 LEMMA.

(i) E _is differentiable of class_ C^∞ _with_

$DE(c)\cdot\eta = \langle \partial c, \nabla\eta\rangle_0$.

(ii) c _is a critical point of_ E _if and only if either_

c = const. **or** c = closed geodesic.

The proof that a closed geodesic c, i.e., a C^∞-curve $c : S \longrightarrow M$ with $\nabla\partial c = 0$, is a critical point follows immediately from (i). To prove the converse one uses the classical Lemma of du Bois-Raymond.

We take on $C^0(s,M)$ the metric

$$d_\infty (c,c') := \sup_{t\in S} d_M(c(t),c'(t)) .$$

On ΛM we have the distance d_Λ derived from the riemannian metric on ΛM .

Then one has

1.11 PROPOSITION. <u>Let</u> $c, c' \in \Lambda M$. <u>Then</u>

$$d_\infty^2(c,c') \le 2d_\Lambda^2(c,c') \quad .$$

<u>Moreover</u>:

$$|\sqrt{2E(c)} - \sqrt{2E(c')}| \le d_\Lambda(c,c')$$

As a consequence it follows

1.12 LEMMA. <u>The inclusion</u>

$$\Lambda(M) = H^1(S,M) \hookrightarrow C^0(S,M)$$

<u>is continuous and compact</u>, <u>i.e.</u>, <u>the image of every bounded set has compact closure</u>.

This is proved using the theorem of Arzela-Ascoli.

As a first consequence we have

1.13 THEOREM. ΛM <u>is a complete metric space with respect to</u> d_Λ . For every real $\kappa \ge 0$ we denote by $\Lambda^\kappa = \Lambda^\kappa M$ the set $c \in \Lambda = \Lambda M$ with $E(c) \le \kappa$. $\Lambda^{\kappa-}$ denotes the interior of Λ^κ .

In particular, $\Lambda^0 = \Lambda^0 M$ is the subset of constant maps $S \to M$. It can be identified with M . We have

1.14 LEMMA. *The natural map*

$$i : M \longrightarrow \Lambda M$$

is an isometric totally geodesic embedding.

1.15 THEOREM. $\Lambda M = H^1(S,M)$ *and* $C^0(S,M)$ *are homotopy equivalent. Actually, the inclusion is a homotopy equivalence.*

1.16 COROLLARY. *The homotopy type of* ΛM *depends only on the homotopy type of* M .

References

[1.1] H. Eliasson, Morse theory for closed curves, Symposium for inf. dim. Topology, Louisiana State University: Ann. of Math. Studies 69 (1972), 63-77. Princeton Univ. Press.

[1.2] P. Flaschel - W. Klingenberg, Riemannsche Hilbertmannigfaltigkeiten. Periodische Geodätische. (Mit einem Anhang von H. Karcher). Lecture Notes in Math. 282 (1972). Springer-Verlag.

[1.3] W. Klingenberg, Closed Geodesics, Ann. of Math. 89 (1969 68-91.

[1.4] R. Palais, Morse theory on Hilbert manifolds, Topology 2 (1963), 299-340.

[1.5] J. Schwartz, Nonlinear functional analysis, with an additional chapter by H. Karcher. Gordon and Breach, New York-London-Paris, 1969.

2. The Morse-Lusternik-Schnirelman-theory on ΛM

Using the riemannian metric $\langle\ ,\ \rangle_1$ on ΛM , we define grad E by

$$\langle \text{grad } E(c), \eta \rangle_1 \dot{=} DE(c) \cdot \eta$$

for all $\eta \in T_c \Lambda$.

The following property of the vector field grad E on ΛM is essential for extending the theory of a differentiable functio[n] on a finite dimensional manifold to the Hilbert manifold ΛM , as was shown by Palais and Smale:

2.1 THEOREM. *The vector field* grad E *satisfies the followi[ng] condition:*

Let $\{c_m\}$ *be a sequence on* ΛM *such that* $\{E(c_m)\}$ *is bounded and* $\lim \| \text{grad } E(c_m) \|_1 = 0$. *Then* $\{c_m\}$ *has an accumulation point* c . *Any such* c *is a critical point of* E .

We denote by $Cr\Lambda$ the set of critical points of Λ . Then $Cr\Lambda \cap \Lambda^{\kappa}$ is a compact set.

2.2 PROPOSITION. *Let* $\kappa > 0$, $K = Cr\Lambda \cap E^{-1}(\kappa)$. *Let* U *be a neighborhood of* K *in* Λ .

Claim: There exists $\varepsilon = \varepsilon(K,U) > 0$ and $\eta = \eta(\varepsilon) > 0$ such that $c \in (\Lambda^{\kappa+\varepsilon} - \Lambda^{(\kappa-\varepsilon)-}) \cap U$ implies:

$$\| \text{grad } E(c) \|_1 \geq \eta$$

Denote by $\phi_s c$ the integral curve of the vector field $\xi = - \text{grad } E$ with $\phi_0 c = c$.

2.3. LEMMA

(i) $\frac{d}{ds} E(\phi_s c) = - \| \xi(\phi_s c) \|_1^2$, hence

$$E(\phi_{s_1} c) - E(\phi_{s_0} c) \leq 0 \quad \text{for} \quad s_0 \leq s_1$$

(ii) $$d_\Lambda^2(\phi_{s_1} c, \phi_{s_0} c) \leq |E(\phi_{s_1} c) - E(\phi_{s_0} c)| |s_1 - s_0|$$
$$\leq E(c) |s_1 - s_0|$$

As a consequence we have:

2.4 THEOREM.

(i) $\phi_s : \Lambda \to \Lambda$ is defined for all $s \geq 0$.

(ii) If $\kappa > 0$ is not a critical value then there exists $\varepsilon > 0$ and $s_0 > 0$ such that

$$\phi_s \Lambda^{\kappa+\varepsilon} \subset \Lambda^{(\kappa-\varepsilon)-}$$

for all $s \geq s_0$.

The following proposition describes the structure of ΛM in a neighborhood of $\Lambda^0 M$.

2.5 PROPOSITION. $\Lambda^0 M$ *is a non-degenerate critical submanifold of* ΛM . *For any sufficiently small* $\kappa > 0$ *is* $\Lambda^0 M$ *a strong deformation retract of* $\Lambda^\kappa M$, *under the map* $\phi_s \Lambda^\kappa \to \Lambda^\kappa, s \to \infty$.

The Lusternik-Schnirelman theory of $\{\Lambda M, -\text{grad } E\}$ deals with the existence of critical points:

2.6 DEFINITION. A non-empty set $\mathcal{A}$ of non-empty subsets A of Λ is called ϕ-*family*, if an

(i) $A \in \mathcal{A}$ implies that $E|A$ is bounded

(ii) $A \in \mathcal{A}$ implies : $\phi_s A \in \mathcal{A}$, for all $s \geq 0$.

2.7 EXAMPLES.

(i) An orbit $\{\phi_s c, s \geq 0\}$

(ii) All elements in a connected component of Λ

(iii) Denote by $[S^k, \Lambda]$ a free homotopy class of maps $S^k \to \Lambda$. Then $\mathcal{A}$:= set of images of this class is a ϕ-family.

(iv) Let z be a singular homology class of Λ or of $\Lambda \bmod \Lambda^0$. Then the set of carriers $|u|$ of cycles $u \in z$ is a ϕ-family.

2.8 DEFINITION. The <u>critical value</u> of a ϕ-family $\mathcal{A}$ is defined as

$$\kappa_{\mathcal{A}} := \inf_{A \in \mathcal{A}} \sup_{c \in A} E(c)$$

2.9 THEOREM. <u>Let</u> $\kappa = \kappa_{\mathcal{A}}$ <u>be the critical value of a</u> ϕ-<u>family</u> $\mathcal{A}$. <u>Then there exists a critical point</u> c <u>with</u> $E(c) = \kappa$.

As a first consequence we have

2.10 THEOREM. <u>Let</u> Λ' <u>be a connected component of</u> Λ <u>which does not contain</u> Λ^0. <u>Then</u> E <u>assumes its infimum on</u> Λ' <u>and any</u> $c \in \Lambda'$ <u>with</u> $E(c) = \inf E|\Lambda'$ <u>is a closed geodesic</u>.

To show the existence of closed geodesics in the case $\pi_1 M = 0$ i.e., ΛM connected, we observe

2.11 LEMMA. <u>The map</u>

$$\gamma : \Lambda M \to M : c \mapsto c(0) = c(1)$$

is a Serre fibration with section $M \to \Lambda^0 M \subset \Lambda M$. *The fibre is the loop space* ΩM .

It follows, for $\pi_1 M = 0$, that

$$\pi_k \Lambda M = \pi_k M + \pi_k \Omega M = \pi_k M + \pi_{k+1} M$$

This now yields the following result of Fet and Lusternik.

2.12 THEOREM. *Let* $\pi_1 M = 0$. *Then there exists on* M *a non-trivial closed geodesic*.

There is a first $k+1$, $1 < k+1 \leq n$, with $\pi_{k+1} M \neq 0$; hence $\pi_k \Lambda M = \pi_{k+1} M \neq 0$. The critical value κ of the ϕ-family given by a non-trivial homotopy class $[S^k, \Lambda M]$, cf. (2.7), is > 0 . Indeed, otherwise there would be for every $\kappa > 0$ an element in this class with $\mathrm{im}(S^k) \subset \Lambda^\kappa$. From (2.5) then follows, that there is a map $S^k \to \Lambda^0 M \cong M$ in this class, which is $= 0$ - a contradiction. To carry over Morse theory one must assume that the critical sets are non-degenerate. We first observe

2.13 LEMMA.

(i) *On* ΛM *there is a continuous* S^1-*action*

$$S^1 \times \Lambda M \longrightarrow \Lambda M$$

given by:

$$(s,c) = (s,c(t)) \longmapsto s \cdot c = (c(t+s)) \quad .$$

Moreover, there is on ΛM an involution

$$\theta : \Lambda M \longrightarrow \Lambda M : c = (c(t)) \longrightarrow \theta c = (c(1-t))$$

These operations are related by

$$s\theta = \theta s^{-1}$$

(ii) The maps $s : \Lambda \to \Lambda : c \to s.c$ and $\theta : \Lambda \to \Lambda : c \to \theta c$ are isometries leaving E invariant. In particular, these operations commute will the E -decreasing deformations $\phi_s : \Lambda \to \Lambda$, $s \in \mathbb{R}$, of (2.3).

In other words, the theory of the space ΛM with the function E is equivariant with respect to the preceding actions of S^1 and $\mathbb{Z}_2$.

In particular, if we want to carry over the Morse theory of a differentiable function to our situation, we cannot assume that the critical points c are non-degenerate because with c also the whole orbit $S.c$ will consist of critical points. What we can assume instead is that, for $c \notin \Lambda^o$, $S.c$ is a non-degenerate critical submanifold in the sense of Bott.

Note that we did show already that the critical submanifold $\Lambda^o M$ of ΛM is always non-degenerate. The prinicpal result of the Morse theory on ΛM for the gradient field of the function E is contained in the following theorem.

We recall that a (connected) compact submanifold B of critical points of ΛM is called <u>non-degenerate</u> if for $c \in B$ the nullity of c is equal to the dimension of B. We denote by $\mu = \mu(B)$ the negative bundle over B, i.e., the i-dimensional vector bundle where the fibre over $c \in B$ is generated by the eigen vectors with negative eigen value. On $\bar{\mu}$ operates S^1. Moreover, $\theta\mu(B) = \mu(\theta B)$.

2.14. THEOREM. <u>Let</u> $\kappa > 0$ <u>be a critical value and assume that the critical points in</u> $E^{-1}(\kappa)$ <u>form non-degenrate critical submanifolds</u> $B, B', \ldots, \theta B, \theta B' \ldots$. <u>Then there exist real numbers</u> $\kappa', \kappa'' : 0 < \kappa' < \kappa < \kappa''$, <u>such that</u> $\Lambda^{\kappa''}$ <u>possesses as strong deformation retract the space</u>.

$$\Lambda^{\kappa'} \underset{h}{\cup} \mu(B) \underset{\theta h}{\cup} \mu(\theta B) \underset{h'}{\cup} \mu(B') \underset{\theta h'}{\cup} \mu(\theta B') \ldots$$

<u>which is obtained from</u> $\Lambda^{\kappa'}$ <u>by attaching equivariantly by negative disc bundles over</u> $B, B', \ldots, \theta B, \theta B'$.

2.15 COMPLEMENT. <u>If all closed geodesics are non-degenrate then,</u>

for each $c \in \Lambda M$, the integral curve $\phi_s(c)$, of the vector-field $-\mathrm{grad}\, E$ starting at c has precisely one limit element $c_o = \lim_{s\to\infty} \phi_s c$ · c_o is a critical element.

References

[2.2] H. Eliasson, Convergence of gradient curves on Hilbert manifolds, Preprint Reykjavik 1973. Submitted to Math. Zeitschrift.

[2.1] R. Bott, Non-degenerate critical manifolds, Ann. of Math. 60 (1954), 248-261.

[2.3] L. Lusternik, The topology of function spaces and the calculus of variations in the large, Trudy. Mat. Inst. Steklov 19 (1947), (Russian).-Translations of Math. Monographs Vol. 16 (1966), Amer. Math. Soc., Providence, R.I.

[2.4] L.A. Lyusternik-A.I. Fet, Variational problems on closed manifolds, Dikl. Akad. Nauk SSSR (N.S.) 81 (1951), 17-18 (Russian).

[2.5] M. Morse, The calculus of variations in the large, Amer. Math. Soc. Colloq. Pub., 18, Amer. Math. Soc. Providence, R.I. 1934.

3. The geodesic flow

Recall the concept of a Hamiltonian system. This is an even-dimensional differentiable manifold N endowed with a symplectic form α , i.e., α is a closed 2-form satisfying $\alpha^m \neq 0$ where $2m = \dim N$. Moreover, there is given a differentiable function $H : N \to R$. H is called Hamilton function and the corresponding Hamilton vector field $\mathfrak{J}_H$ is defined by

$$\mathfrak{J}_H \alpha = dH$$

Given a differentiable manifold M , the cotangent bundle T^*M carries a canonical symplectic structure $\alpha = -d\theta$, with $\theta = \sum_i y_i dx^i$, where (x^i) are coordinates on M and (x^i, y_i) the corresponding canonical coordinates on T^*M .

If we have a riemannian metric on M then this defines an isomorphism between Tm and T^*M , which is given in local coordinates by

$$\sum_i \xi^i(x) \frac{\partial}{\partial x^i} \longrightarrow \sum_{i,k} \xi^i(x) g_{ik}(x) dx^k$$

If we now take the kinetic energy

$$X = \sum_i \xi^i \frac{\partial}{\partial x^i} \longmapsto \frac{1}{2} \langle X,X \rangle = \frac{1}{2} \sum_{i,k} g_{ik} \xi^i \xi^k$$

we obtain a Hamilton system on TM which is called the <u>geodesic spray</u>.

In the canonical local coordinates $(x^i, \dot{x}^i)$, of TM associated to local coordinates (x^i) of M, the geodesic spray is given by

$$\frac{d}{dt}x^i = \dot{x}^i \ , \ \frac{d}{dt}\dot{x}^i = -\sum_{k,l} \Gamma^i_{kl}(x)\dot{x}^k\dot{x}^l$$

where $\Gamma^i_{kl}(x)$ are the Christoffel symbols.

We see that the integral curves of the geodesic spray are the tangent vector fields $\dot{c}(t)$ of geodesics $c(t)$ on M. We recall that the tangent bundle TTM of the tangent bundle TM of a riemannian manifold M has a canonical decomposition:

$$TTM = T_hTM \oplus T_vTM$$

into the so-called <u>horizontal</u> and <u>vertical</u> subbundles: T_vTM is the bundle tangent to the fibre and T_hTM is given by the Levi-Civitè connection as follows: If $(x, \dot{x}, \xi, \eta)$ are the canonical coordinates of TTM associated to the coordinates $(x, \dot{x})$ of TM then the vertical part is given by $(x, \dot{x}, 0, \eta)$ and the horizontal part is given by $(x, \dot{x}, \xi, -\Gamma(x)(\dot{x}, \xi))$ with $\Gamma^i(x)(\dot{x}, \xi) = \sum_{k,l} \Gamma^i_{kl}(x)\dot{x}^k\xi^l$. The horizontal subspace $T_{Xh}TM$ and the vertical subspace $T_{Xv}TM$ are canonically identified with the tangent space T_pM at $p = \tau X \in M$ by

$$K : (x,\dot{x},\xi,-\Gamma(x)(\dot{x},\xi)) \longmapsto (x,\xi)$$

$$d\tau : (x,\dot{x}.0,\eta) \longmapsto (x,\eta)$$

It follows that hte vector of the geodesic spray at $X \in TM$ is given by $(X,0) \in T_{Xh} \oplus T_{Xv}$.

The integral curves $\phi_t : TM \to TM$ of the geodesic spray define the <u>goedesic flow</u> on TM . The projection $\tau\phi_t X$ of an orbit $\phi_t X$ into M is the geodesic $c(t)$ determined by $\dot{c}(0) = X$.

We recall that a <u>Jacobi field</u> $Y(t)$ along a geodesic $c(t)$ is a vector field satisfying

$$\nabla^2 Y(t) + R(\dot{c}(t),Y(t))\dot{c}(t) = 0$$

where $R(\ \ ,\ \ )$ is the curvature tensor.

3.1 PROPOSITION. <u>The invariant vector fields</u> $\tilde{Y}(t)$ <u>along an orbit</u> $\phi_t X_o$ <u>of the geodesic flow are given by</u>

$$\tilde{Y}(t) = (Y(t),\nabla Y(t)) \in T_{\phi_t X_o h} \oplus T_{\phi_t X_o v}$$

<u>where</u> $Y(t)$ <u>is a Jacobi field along the underlying geodesic</u>

$c(t) = \tau \circ \phi_t X_o$.

Assume: $\dim M = n+1$. T_1M shall denote the unit tangent bundle over M . T_1M is left invariant under ϕ_t . We denote by

$$\tau_h^n : T_h^n T_1 M \longrightarrow T_1 M$$

$$\tau_v^n : T_v^n T_1 M \longrightarrow T_1 M$$

the subbundles of TT_1M which are given by the horizontal and vertical vectors respectively which are orthogonal to the geodesic spray $(X_o,0)$ at $X_o \in T_1M$.

3.2 PROPOSITION. _Let_ $\phi_t X_o$ _be a periodic orbit of period_ $\omega > 0$, $|X_o| = 1$. _Put_ $\tau\phi_t X_o = c(t)$. _Let_ Σ _be a local hypersurface in_ T_1M , _orthogonal to_ $(X_o,0)$ _in_ X_o . Σ _has dimension_ $2n$ _and it carries a natural symplectic structure induced from the symplectic structure on_ TM .

Claim: _There are open neighborhoods_ Σ_o _and_ Σ_ω _of_ X_o _on_ Σ _and a differentiable function_.

$$\delta : \Sigma_o \longrightarrow R \quad \text{with} \quad \delta(X_o) = 0$$

such that

$$X \in \Sigma_o \longmapsto \phi_{\omega+\delta(X)} X \in \Sigma_\omega$$

is a symplectic diffeomorphism $\mathcal{P}_c$. *Moreover, for* $0 < t < \omega + \delta(X)$, $\phi_t X$ *does not belong to* Σ .

$\mathcal{P}_c$ is called *Poincaré map* (associated to c).

The linear approximation of $\mathcal{P}_c$ in X_o , i.e., $d\mathcal{P}_{X_o} : T_{X_o}\Sigma \to T_{X_o}\Sigma$ is also called (*linear*) *Poincaré* map.

The importance of the Poincaré map $\mathcal{P}_c$ lies in the fact that periodic points of $\mathcal{P}_c$ are in 1:1 correspondence with periodic orbits near $\phi_t X_o$ and hence near $c(t) = \tau \circ \phi_t X_o$.

3.3 DEFINITION. A periodic orbit $\phi_t X_o$ (or the underlying closed geodesic $c(t)$) is called *hyperbolic* if the linear symplectic map P_c has no eigenvalue of modulus 1. If all eigenvalues of P_c have the modulus 1, c is called *elliptic*.

3.4. PROPOSITION. *If the sectional curvature* K *of* M *is strictly negative along a closed geodesic* c *then* c *is hyperbolic*. Note, however, that also for $K > 0$ a periodic orbit of ϕ_t can be hyperbolic.

For example, on an ellipsoid with three pairwise different axis the ellipse in the plane through the shortest and the longest axis is hyperbolic; this is also the axis which contains the four umbilics. The other two ellipses on the ellipsoid are elliptic, however.

3.5. PROPOSITION. *If* c *is hyperbolic then* $\mathcal{P}_c$ *has no periodic points* $X \neq X_o$ *near* X_o. *More precisely, there exist in* $\Sigma = \Sigma^{2n}$, *for* Σ *sufficiently small, the stable and the unstable submanifold* Σ_s^n *and* Σ_u^n *which are invariant under* $\mathcal{P}_c$ *such that* $\mathcal{P}_c|\Sigma_s^n$ *and* $\mathcal{P}_c|\Sigma_u^n$ *are contracting and expanding maps respectively.*

A first relation between a property of a closed geodesic c , considered as critical point of ΛM , and a property of the corresponding periodic orbit $\dot{c}(t) = \phi_t \dot{c}(0)$ in TM is given by the following

3.6 LEMMA. *A closed geodesic* c *is non-degenerate (in the sense that the orbit* $S.c$ *is a non-degenerate critical submanifold of* ΛM) *if and only if the Poincaré map* P_c *has no eigenvalue equal to* 1.

W. Klingenberg

References:

[3.1] R. Abraham - J. Marsden, Foundations of Mechanics, Benjamin, New York and Amsterdam, 1967.

[3.2] V.I. Arnold - A. Avez, Problèmes ergodiques de la mèchanique classique, Gauthier-Villars, Paris 1967.

[3.3] W. Klingenberg, Hyperbolic closed geodesics, pp. 155-164 in: Dynamical systems, ed. M.M. Peixoto, Acad, Press New York - London 1973.

[3.4] W. Klingenberg, Manifolds with geodesic flow of Anosov type, Ann. of Math. 99 (1974), 1-13.

[3.5] J. Moser, Stable and random motions in dynamical systems Ann. of Math. Studies 77, Princeton University Press, Princeton, N.Y., 1973.

4. The index theorem for closed geodesics.

Let $c(t)$, $0 \le t \le 1, |\dot{c}(t)| = \omega > 0$, be a closed geodesic of a riemannian manifold M . As we have seen in (2), c can be viewed as a critical point of the energy integral E on the Hilbert manifold $\Lambda = \Lambda M$. As such, there is associated to c the Hessian $D^2E(c)$.

4.1 LEMMA. $D^2E(c)$ *is a quadratic form on the Hilbert space* $T_c\Lambda$, *given by*

$$D^2E(c)(\xi,\xi') = <\nabla\xi,\nabla\xi'>_0 + <R_c(\xi),\xi'>_0$$

$$<\xi,\xi'>_1 + <(R_c-id)\xi,\xi'>_0$$

with $$R_c(\xi)(t) = R(\dot{c}(t),\xi(t))\dot{c}(t)$$

The corresponding selfadjoint operator A_c :

$$<A_c\xi,\xi'>_1 = D^2E(c)<\xi,\xi'>$$

is of the form

$$A_c = id + k_c$$

where $k_c = (id-\nabla^2)^{-1} \circ (R_c-id)$ *is a compact operator.*

As a consequence, the eigenvalues of A_c are discrete with limit value 1.

Index c is the dimension of the subspace of $T_c\Lambda$ spanned by the eigenvectors belonging to negative eigenvalues.

We know that a closed geodesic c also determines a periodic orbit $\phi_t X_o, 0 \le t \le \omega$, of the geodesic flow with $X_o = \dot{c}(0)/\omega$.

We want to give a description of index c using the second interpretation. To do this we observe that the immersion

$$S_\omega \longrightarrow T_1M : t \longmapsto \phi_t X_o$$

induces a 2n-dimensional vector bundle

$$\tau^{2n} : V^{2n} \longrightarrow S_\omega$$

where the fibre $V^{2n}(t)$ over $t \in S_\omega$ consists of the vectors $\tilde{X} \in T_{\phi_t X_o} T_1M$ orthogonal to the geodesic spray $(\phi_t X_o, 0)$. Here we have put $\dim M = n+1$.

The symplectic structure on TM induces a symplectic structure on $V^{2n}(t)$ which is given by

$$2\alpha((X_h, X_v), Y_h, Y_v)) = \langle X_h, Y_v \rangle - \langle Y_h, X_v \rangle$$

This structure is kept invariant under the geodesic flow ϕ_t induced on τ^{2n}, i.e., we have using (3.1):

$$\langle Y(t), \nabla Z(t)\rangle - \langle Z(t), \nabla Y(t)\rangle = \text{const}$$

for Jacobi fields $Y(t), Z(t)$ along $\tau\phi_t X_o$.

In particular, $d\phi_\omega : V^{2n}(0) \to V^{2n}(0)$ is the linear Poincaré map P_c introduced in (3).

We need a certain normal form of a linear real symplectic map $P : V \to V$.

To describe such a normal form we complexify $V : V \otimes \mathbb{C}$, and write again V . The skew symmetric form α on V shall be extended to a skew hermitean form and P shall commute with multiplication by complex numbers.

If ρ is an eigen value of P , also $\bar{\rho}, \rho^{-1}\ \bar{\rho}^{-1}$ are eigen values of P . We denote by $V(\rho)$ the generalized eigen space belonging to the eigen value ρ , i.e., $V(\rho)$ consists of the elements $X \in V$ with $(P\rho^{-1}-1)^k X = 0$ for some integer $k \geq 0$ . .

4.2 THEOREM. For a symplectic map

$$P : V \to V$$

there exists an orthogonal decomposition

$$V = V_{nc} \oplus V_{co}$$

determined up to a symplectic isomorphism, into non-degenerate invariant subspaces of dimension $2p_{nc}$ and $2p_{co}$ respectively such that $P|V_{nc}$ is non-compact and $P|V_{co}$ is compact.

Moreover, V_{nc} has a decomposition

$$V_{nc} = V_{su} \oplus V_{od} \oplus V_{ev}$$

into non-degenerate invariant subspaces of dimension $2p_{su}$, $2p_{od}$ and $4p_{ev} + 2k_{ev}$ with the following properties:

(a) V_{su} is the direct sum of non-degenerate subspaces of the form

$$V(\rho) \oplus V(\bar{\rho}^{-1})$$

ρ an eigen value of P with $|\rho| \neq 1$.

(b) V_{od} is the direct sum of non-degenerate invariant subspaces $V_{od,\rho}$, ρ an eigen value of P with $|\rho| = 1$, which has a base of the form

$$(P\bar{\rho}-1)^j X, \quad 0 \leq j \leq 2l-1$$

with

$$\alpha((P\bar{\rho}-1)^l X,(P\rho-1)^l X) = \pm 1$$

with well determined sign.

(c) V_{ev} is direct sum of non-degenerate invariant subspaces

$$V_{ev,\rho,\bar{\rho}} = V_{ev,\rho} \oplus \bar{V}_{ev,\rho}$$

where ρ is an eigen value of P with $|\rho| = 1$ and $V_{ev,\rho}$ has a base of the form

$$(P\bar{\rho}-1)^j X, \quad 0 \leq j \leq 2l, \quad l > 0$$

with

$$\alpha((P\bar{\rho}-1)^l X,(P\rho-1)^l \bar{X}) = \pm i$$

with well determined sign. k_{ev} is the number of such subspaces.

4.3 COMPLEMENT

(i) $V_{sn} = V_{su}^{2p_{su}}$ has the subspace $V_s = V_s^{p_{su}}$ as invariant isotropic subspace which is formed by the $V(\rho)$ with $|\rho| > 1$.

(ii) $V_{od} = V_{od}^{2p_{od}}$ has the space $V_{od,in} = V_{od,in}^{p_{od}}$ as invariant isotropic subspace which is formed by the subspaces $V_{od,in,\rho} \subset V_{od}$, having a base $(P\bar{\rho}-1)^j X$, $1 \leq j \leq 2l-1$.

(iii) $V_{ev} = V_{ev}^{4p_{ev}+2k_{ev}}$ has the space $V_{ev,in} = V_{ev,in}^{2p_{ev}}$ as

<u>invariant isotropic subspace which is formed by the subspaces</u>

$$V_{ev,in,\rho,\bar{\rho}} = V_{ev,in,\rho} \oplus \bar{V}_{ev,in,\rho}$$

<u>having a base</u>

$$(P\bar{\rho}-1)^j X, (P\rho-1)^j \bar{X}, \; 1+1 \le j \le 2l \quad .$$

Using the preceding decomposition we have with

$$V^p_{in} = V_s \oplus V_{od,in} \oplus V_{ev,in}$$

a real invariant isotropic subspace of dimension
$p = p_{su} + p_{od} + p_{ev}$.

For each subspace $V_{ev,\rho,\bar{\rho}}$ of the type described under c) we define a 2-dimensional real non-degenrate subspace $V^2_{ev,un,\rho,\bar{\rho}}$ as the space generated by $(P\bar{\rho}-1)^l X$, $(P\rho-1)^l \bar{X}$. The direct sum of these spaces, taken over all eigen values ρ , shall be denoted by $V_{ev,un} = V^{2k_{ev}}_{ev,un}$.

We put

$$V^{2q}_{un} = V^{2k_{ev}}_{ev,un} \oplus V^{2q_{co}}_{co}$$

The orthogonal complement of V^{2q}_{un} in V^{2n} will be denoted by

$V_{in}^{2p} \cdot V_{in}^{2p}$ contains the invariant isotropic subspace V_{in}^{p} defined above.

4.4 PROPOSITION. *Let*

$$V^{2n} = V_{in}^{2p} \oplus V_{un}^{2q}$$

be an orthogonal decomposition into non-degenerate real subspaces. Then the projection of

$$V_v^n \cap (V_{in}^p \oplus V_{un}^{2q})$$

into V_{un}^{2q} *modulo* V_{in}^{p} *yields a real isotropic subspace* V_{un}^{q} $V_e^n := V_{in}^p \oplus V_{un}^q$ *is isotropic and there exists a q-dimensional subspace* $V_v^q \subset V_v^n$ *with*

$$V_e^n = V_{in}^p \oplus V_v^q$$

To formulate now our index theorem we first define the ρ-*index*: Let $c = c(t)$, $0 \le t \le \omega$, $|\dot{c}(t)| = 1$, be a closed geodesic. Let $\rho \in S^1 \subset \mathbb{C}$. Denote by $\rho T_c = \rho T_c \Lambda M$ the Hilbert space of complex valued vector field $\mathfrak{Y}(t)$ along $c(t)$ satisfying $\bar{\rho}\mathfrak{Y}(\omega) = \mathfrak{Y}(0)$, endowed with the scalar product

$$\langle \mathfrak{Y}, \mathfrak{Y}' \rangle_1 := \langle \nabla \mathfrak{Y}, \overline{\nabla \mathfrak{Y}'} \rangle_0 + \langle \mathfrak{Y}, \bar{\mathfrak{Y}}' \rangle_0 \; .$$

Clearly, also the form $D^2E(c)$ can be extended to a hermitian form $D^2E\rho(c)$ on $\rho T_c\Lambda$.

4.5 DEFINITION. The ρ-*index of* c , $I_c(\rho)$, is defined as the dimension of the subspace of $\rho T_c\Lambda$ generated by the eigen vectors of $D^2E\rho(c)$ having negative eigen values.

Finally, we define on the isotropic subspace $V^q_{un} \subset V^{2q}_{un}$ for each $\rho \in S^1 \subset$ which is not an eigen value of P , a bilinear form

$$2\rho(X,Y) := -2\alpha(X,(P\rho-1)^{-1}\bar{Y})$$

4.6 PROPOSITION. Q_ρ *is a hermitian form on* V^q_{un} . We now can formulate the ρ-index theorem:

4.7 THEOREM. *Let* $\rho \in \mathbb{C}$, $|\rho| = 1$, *not be eigen value of the Poincaré map* P_c *of a closed geodesic* c . *Then the* ρ-*index of* c *is given by*

$$I_c(\rho) = J_c + (\text{index} + \text{nullity})\, Q\rho$$

where

$$J_c = \sum_{0<t<\omega} \dim V^n_v(t) \cap d\phi_t V^n_e + \dim(V^n_v(0) \cap V^p_{in}) \quad .$$

If $\rho = \rho_0$ is an eigen-value of P_c , we define the *splitting* numbers $S_c^{\pm}(\rho_0)$ as follows:

4.8 DEFINITION.

$$S_c^{+}(\rho_0) := \lim_{\rho \to \rho_0} (I_c(\rho) - I_c(\rho_0)) \ , \quad \arg \rho\bar{\rho}_0 > 0 \ .$$

$$S_c^{-}(\rho_0) := \lim_{\rho \to \rho_0} (I_c(\rho) - I_c(\rho_0)) \ , \quad \arg \rho\bar{\rho}_0 < 0 \ .$$

These numbers are determined by P_c alone. We can give an explicit description using the normal form (4.3) of P_c :

1. Let $V_{co}(\rho_0, \bar{\rho}_0) \subset V_{co}$ be the subspace spanned by the eigen vectors with eigen-value ρ_0 and $\bar{\rho}_0$. Put

$$\dim V_{co}(\rho_0, \bar{\rho}_0) = 2k_{co}(\rho_0, \bar{\rho}_0) \ .$$

If $\rho_0^2 \neq 1$, denote by $k_{co}^{+}(\rho_0, \bar{\rho}_0)$ and $k_{co}^{-}(\rho_0, \bar{\rho}_0)$ the number of 2-dimensional invariant real subspaces with a base $X_0, \bar{X}_0$ satisfying

$$\alpha(X_0, \bar{X}_0) = +i \quad \text{or} \quad -i \ ,$$

respectively.

2. Let $V_{ev,un}(\rho_0, \bar{\rho}_0)$ $V_{ev,un}$ be the subspace formed by 2-dimensional real spaces having a base of the form

$$(P\bar{\rho}_0-1)^l X,\ (P\rho_0-1)^l \bar{X}$$

with

$$\alpha((P\bar{\rho}_0-1)^l X,\ (P\rho_0-1)^l \bar{X}) = \pm i$$

Put $\dim V_{ev,un}(\rho_0,\bar{\rho}_0) = 2k_{ev}(\rho_0,\bar{\rho}_0)$. Denote by $k^+_{ev}(\rho_0,\bar{\rho}_0)$ and $k^-_{ev}(\rho_0,\bar{\rho}_0)$ the number of such 2-dimensional spaces for which in the above equation $+i$ and $-i$ stands, respectively.

3. Denote by $k^-_{od}(\rho_0,\bar{\rho}_0)$ the number of invariant subspaces of V_{od} with a base of the form

$$(P\bar{\rho}_0-1)^j X,\ (P\rho_0-1)^j \bar{X},\ 0 \le j \le 2l-1 \quad (\text{if } \rho_0^2 \neq 1)$$

or

$$(P\bar{\rho}_0-1)^j X,\ 0 \le j \le 2l-1 \quad (\text{if } \rho_0^2 = 1)$$

for which we have

$$\alpha((P\bar{\rho}_0-1)^l X,\ (P\rho_0-1)^{l-1} \bar{X}) = -1$$

Using these notations we formulate the following

4.9 THEOREM. <u>The splitting numbers</u> $S^\pm_c(\rho_0)$ <u>of a closed</u>

<u>geodesic</u> c <u>at the eigen-value</u> ρ_0, $|\rho_0| = 1$, <u>of</u> P_c <u>are given as follows</u>: <u>If</u> $\rho_0^2 \neq 1$:

$$S_c^{\pm}(\rho_0) = k_{od}^-(\rho_0,\bar{\rho}_0) + k_{ev}^{\pm}(\rho_0,\bar{\rho}_0) + k_{co}^{\pm}(\rho_0,\bar{\rho}_0)$$

<u>If</u> $\rho_0^2 = 1$:

$$S_c^{\pm}(\rho_0) = k_{od}^-(\rho_0,\bar{\rho}_0) + k_{ev}(\rho_0,\bar{\rho}_0) + k_{co}(\rho_0,\bar{\rho}_0)$$

The importance of the ρ-index of c lies in the fact that it allows to determine the index of all coverings c^m of c :

4.10 LEMMA. <u>Let</u> c <u>be a closed geodesic</u>.

<u>Then</u>

$$\text{index } c^m = \sum_{\rho^m=1} I_c(\rho) \quad .$$

<u>The function</u> $\rho \in S^1 \to I_c(\rho) \in N$ <u>is up to a constant determined by the Poincaré map</u> P_c . <u>The constant is determined by</u> (4.7).

Note that in the special case that c is hyperbolic, i.e., where P_c has no eigen-values ρ_0 with $|\rho_0| = 1$, the previous results become considerably more simple. In this case, V^{2n} has an invariant isotropic subspace of maximal dimension: $V_{in}^n = V_s^n$. Hence:

$$\text{index } c = \sum_{o \leq t \leq \omega} \dim V_v^n(t) \cap d\phi_t V_{in}^n$$

$$\text{index } c^m = m \cdot \text{index } c \quad .$$

References:

[4.1] V.I. Arnold, Characteristic class entering in quantization conditions, Funkcional. Anal. i Prolzen 1 (1967) 1-14. (russ.); Funkcional Anal. Appl. 1 (1967) 1-13.

[4.2] R. Bott, On the iteration of closed geodesics and the Sturm intersection theory, Comm. Pure Appl. Math. 9 (1956), 171-206.

[4.3] H. Duistermaat, On the Morse Index in Variational Calculus. Preprint, Nijmwegen 1974.

[4.4] W. Klingenberg, Le théorème de l'indice pour les géodésiques fermées, C.R. Acad. Sc. Paris 276 (1973), 1005-1009.

[4.5] W. Klingenberg, The index theorem for closed geodesics, Tohoku Math. J. (to appear).

[4.6] W. Klingenberg, Der Indexsatz for geschlossene Geodätische, Preprint Bonn 1974.

5. An application of the Birkhoff-Lewis fix point theorem.

Let $c = c(t)$, $0 \leq t \leq \omega$, $|\dot{c}(t)| = 1$, be a closed geodesic. In (3) we have introduced the Poincaré map

$$\mathcal{P}_c : (\Sigma_0, X_0) \longrightarrow (\Sigma_\omega, X_0)$$

associated to c with $X_0 = \dot{c}(0)$.

$\mathcal{P}_c$ is a symplectic diffeomorphism. As we did observe already in (3), the periodic points of $\mathcal{P}_c$, i.e., points X for which there exists a positive inter $N = N(X)$ with $\mathcal{P}_c^N X = X$, corresponding to periodic orbits $\phi_t X$ lying in a tubular neighborhood of $\phi_t X_0$ with a period $\sim N_\omega$. If $\mathcal{P}_c$ (or c) is hyperbolic then $\mathcal{P}_c$ will have no periodic points different from X_0, cf. (3.5). Therefore we consider the partially elliptic case, i.e., we assume that among the eigen-values of P_c there are some of modulus 1. From (4.3) we see that the number of such eigen-values counted with multiplicity is always even, say 2r.

5.1 DEFINITION. The Poincaré map $\mathcal{P}_c$ is called 4-elementary if the 2r eigen-values ρ of P_c with $|\rho| = 1$ satisfy the following condition. If we write these ρ in the form

$$\rho_j = e^{i\alpha_j}, \quad \bar{\rho}_j = e^{-i\alpha_j}, \quad 1 \leq j \leq r \quad \text{and} \quad 0 \leq \alpha_j \leq \pi \text{ , then,}$$

for any family $(k_j)_{1 \leq j \leq r}$ of integers satisfying $1 \leq \sum |k_j| \leq 4$ we shall have $\prod_j \rho_j^{k_j} \; 1$, i.e., $\sum \alpha_j k_j \neq 0 \bmod 2\pi$.

Note that the property "4-elementary" for $\mathfrak{P}_c$ remains invariant for small changes of the riemannian metric on M which leave the geodesic c invariant. We therefore say that "4-elementary" is an "open" property.

That this property is also "dense" says the following

5.1 LEMMA. *Let* $\phi_t X_0$, $0 \leq t \leq \omega$, *be a periodic orbit of the geodesic flow on* $T_1 M$ *and* $c(t) = \tau \phi_t X_0$ *the corresponding closed geodesic* c .

Then there exists arbitrarily C^∞*-close to the riemannian metric* g *a riemannian metric* g^* *such that* $\phi_t^* X_0$ *is again periodic and* $c^*(t) = \tau \phi_t^* X_0$ *is 4-elementary*.

The proof can according to Abraham be given by using a transversality argument. An other proof would use methods of control theory. Before we continue we state an important theorem which allows us to restrict our considerations to Poincaré maps of purely elliptic type:

5.2 THEOREM. *Let* $\phi_t X_0, |X_0| = 1$, *be a periodic orbit of the geodesic flow and assume that there are* $2r > 0$ *eigen values* of modulus 1 for P_c and that:

$$\mathfrak{P}_c : (\Sigma_0, x_0) \longrightarrow (\Sigma_\omega, x_0)$$

is 4-elementary. Hence, in particular, there is no eigen-value equal to 1 and the eigen values ρ with $|\rho| = 1$ have multiplicity 1. Put $n-r = m \geq 0$.

Claim: There exist local submanifolds Σ_s^m, Σ_u^m. Σ_{ce}^{2r} in Σ passing through the origin x_0 of Σ such that $\Sigma_s^m \cap \Sigma_0$, $\Sigma_u^m \cap \Sigma_0$ and $\Sigma_{ce}^{2r} \cap \Sigma_{co}$ are mapped under $\mathfrak{P}_c$ diffeomorphically onto $\Sigma_s^m \cap \Sigma_\omega$, $\Sigma_u^m \cap \Sigma_\omega$ and $\Sigma_{ce}^{2r} \cap \Sigma_\omega$ respectively.

The action of $\mathfrak{P}_c$ on Σ_s^m and Σ_u^m is contracting and expanding, respectively, and of class C^∞ . The action on Σ_{ce}^{2r} is of finite, but arbitrarily high order of differentiability for sufficiently small Σ_{ce} . $\mathfrak{P}_c | T_{x_0} \Sigma_{ce}$ has all eigen-values of modulus 1 .

Σ_s and Σ_u are called stable and unstable manifold, respectively, and Σ_{ce} is called center manifold.

As a consequence, if we want to prove the existence of periodic points we can restrict ourselves to the case of a local symplectic diffeomorphism of class C^s , s large:

$$(*) \qquad \mathfrak{P}: (\mathbb{R}^{2r}, 0) \longrightarrow (\mathbb{R}^{2r}, 0)$$

where the linear part P is 4-elementary. For such a $\mathfrak{P}$ one can find according to Birkhoff a certain normal form:

5.3 PROPOSITION. *Let $\mathfrak{P}$, (*), be a 4-elementary local symplectic diffeomorphism. Then there exist canonical (= symplectic) complex conjugate coordinates* $(z_j, \bar{z}_j)$, $1 \leq j \leq r$, *such that $\mathfrak{P}$ is given by*

$$z_j^* = z_j\, e^{i(\alpha_j + \sum_k \beta_{jk} z_k \bar{z}_k)} + O(z\bar{z})^2$$

where $\rho_j = e^{i\alpha_j}$ *are the eigen-values of* $d\mathfrak{P}(0)$, $0 < \alpha_j < \pi$. *and the* β_{jk} *are real. The matrix* (β_{jk}) *is determined up a change of its rows and columns.*

5.4 DEFINITION. Let $\mathfrak{P}$, (*), be a local 4-elementary symplectic diffeomorphism. $\mathfrak{P}$ is called a *twist map* if in the normal form (5.3) we have $\det(\beta_{jk}) \neq 0$.

For example, if $\mathfrak{P}$ is a local symplectic 4-elementary diffeomorphism of the plane, i.e., $r = 1$, then the normal forms looks like

$$z^* = ze^{i(\alpha + \beta z\bar{z})} + O(z\bar{z})^2$$

and this is a twist map if $\beta \neq 0$.

Such 2-dimensional twist maps are stable, as was shown by

Kolmogoroff, Arnold and Moser.

We are interested in an other property of twist maps, i.e., the so-called Birkhoff-Lewis fix point theorem. In its differentiable version which is due to Moser, Cushman and Marzouk it reads as follows:

5.5 THEOREM. Assume that the local symplectic map $\mathfrak{P}$, (*), is 4-elementary and twist. Then there exist in every neighborhood of the origin an infinite number of periodic points with period going to infinity.

The idea of the proof is that, since $\mathfrak{P}$ is a twist map, the behavior near the origin is dominated by the 3-jet of $\mathfrak{P}$. Now, since $\det(\beta_{jk}) \neq 0$, the set of $z = (z_1, \ldots z_r) \in \mathbb{C}^r$ in a neighborhood of the origin for which the exponents

$$i(\alpha_j + \sum_k \beta_{jk} z_k \bar{z}_k)$$

become a rational multiple of $i2\pi$, say $i2\pi p_j/q_j$ is dense. But if $\mathfrak{P}(z)$ is given by

$$z_j^* = z_j e^{i2\pi p_j/q_j}$$

then there is a power of $\mathfrak{P}$ leaving z invariant, i.e., z is a periodic point.

As a consequence we have

5.6 THEOREM. <u>Let</u> c <u>be a non-hyperbolic closed geodesic on</u> M <u>and assume that the corresponding Poincaré map</u> $\mathcal{P}_c$ <u>is 4-elementary and</u> (<u>restricted to the center manifold</u>) <u>twist</u>. <u>Then there exist in every tubular neighborhood of</u> c <u>an infinite sequence of closed geodesics with length going to infinity</u>.

Given a non-hyperbolic closed geodesic c, the Poincaré map $\mathcal{P}_c$ need not to satisfy the conditions of the theorem (5.6). However, by perturbing the given riemannian metric g arbitrarily little, one can achieve this.

To formulate this more precisely we introduce the concept of a generic property: First, for a given compact differentiable manifold M we denote by $\mathcal{G}(M)$ the set of C^r riemannian metrics, endowed with the strong C^r-topology, $5 \leq r \leq \infty$.

5.7 DEFINITION. A property $\mathcal{E}$ of a riemannian metric g on M is called <u>generic</u> if the set $\mathcal{E}$ of $g \in \mathcal{G}(M)$ satisfying this property contains a residual subset of $\mathcal{G}(M)$, i.e., a set which can be represented as the intersection of a countable number of open dense subsets of $\mathcal{G}(M)$.

Using this concept, one can prove:

5.8 THEOREM. <u>Let</u> M <u>be a compact differentiable manifold. Then the following property on</u> $\mathcal{G}(M)$ <u>is generic: For every non-hyperbolic closed geodesic</u> c <u>on</u> (M,g) <u>the Poincaré map</u> $\mathcal{P}_c$ <u>is 4-elementary and twist.</u>

Note that this will imply, using (5.6), that generically there are infinitely many prime closed geodesics on (M,g) if one excludes that there exists a $g_0 \in \mathcal{G}(M)$ such that (M,g_0) has only finitely many prime closed geodesics all of which are hyperbolic. The same will then be true for (M,g) with g in a small neighborhood of g_0 in $\mathcal{G}(M)$. We will see in (6) that such g_0 can not exist if $\pi_1 M < \infty$.

<u>References</u>:

[5.1] G.D. Birkhoff, Dynamical systems, Amer. Math. Soc. Colloq. pub., vol. 9, Amer. Math. Soc., New York, N.Y., 1927. Revised ed. 1966.

[5.2] G.D. Birkhoff, On the periodic motions near a given periodic motion of a dynamical system, Ann. di Mat (4) 12 (1933), 117-133.

[5.3] I.C. Harris, Periodic solutions of arbitrarily long periods in Hamilton systems, J. Diff. Equations 4 (1968), 131-141.

[5.4] W. Klingenberg - F. Takens, Generic properties of geodesic flows, Math. Ann. <u>197</u> (1972), 323-334.

[5.5] J. Moser, Über periodische Lösungen kanonischer Differentialgleichungssysteme, Nachr. Akad. Wiss. Göttingen, Math. Phys. Chem. Abt. (1953), 23-48.

[5.6] J. Moser, Stabilitätsverhalten kanonischer Differentialgleichungssysteme, Nachr. Akad. Wiss. Göttingen, Math. Phys. Kl. IIa (1955), 87-120.

[5.7] M. Marzouk, Der Fixpunktsatz von Birkhoff-Lewis im differenzierbaren Fall. Dissertation, Bonn 1974.

[5.8] C.L. Siegel and J.K. Moser, Lectures on Celestical Mechanics, Springer Verlag, Berlin-Heidelberg-New York 1971.

[5.9] S. Sternberg, Celestial mechanics, part II, Benjamin, New York 1969.

W. Klingenberg

6. The Existence of many closed geodesics.

For simplicity we consider a simply connected compact riemannian manifold M. If we want to specify the riemannian metric g on M we also write (M,g) instead of M and denote by M the underlying compact differentiable manifold.

We know from (2) that there is always at least one closed geodesic on M which we can assume to be prime. The proof of this theorem of Fet and Lusternik was based on the fact that $\Lambda M \bmod \Lambda^{o} M$ is homologically non-trivial, i.e., if $k+1 \geq 2$ is the first non-vanishing (rational) Betti number of M then also $H_k(\Lambda,\Lambda^{o}) \neq 0$, with $\Lambda = \Lambda M$.

To obtain the existence of more geodesics one can use the theory of subordinated homology classes. We first observe that there is a linear pairing, given by the cap product:

$$\cap : H_*(\Lambda,\Lambda^{o}) \otimes H^*(\Lambda-\Lambda^{o}) \rightarrow H_*(\Lambda,\Lambda^{o})$$

6.1 DEFINITION. Let z', z be non-trivial homology classes of $\Lambda \bmod \Lambda^{o}$, $\dim z > \dim z' > 0$. z' is called subordinated to z if there is a cohomology class $\mathfrak{I}$ of $\Lambda - \Lambda^{o}$ such that $z' = z \cap \mathfrak{I}$. Notation: $z \succ z'$.

6.2 LEMMA. Let

$$z \succ z' \ldots\ldots \succ z^{(r)}$$

be a chain of subordinated homology classes of $\Lambda \bmod \Lambda^{o}$. Then the critical values $\kappa, \kappa', \ldots, \kappa^{(r)}$ satisfy

$$\kappa \geq \kappa' \geq \ldots\ldots \geq \kappa^{(r)} > 0$$

and if equality holds e.g. at the jth place: $\kappa^{(j)} = \kappa^{(j+1)}$, then there is a family of critical points (= closed geodesic) of E-value $\kappa^{(j)}$ having covering dimension $\geq \dim z^{(j)} - \dim z^{(j+1)} > 0$.

This result can be refined by taking into account that the Morse-Lusternik-Schnirelman theory of the function E on ΛM is actually equivariant with respect to the S^1-action on the involution θ on ΛM which we did describe in (2).

We mention the following result:

6.3 THEOREM. Let $\pi_1 M = 0$ and $1 \geq 2$ the first dimension with $H_1(M, Z_2) \neq 0$. Consider a map

$$(*) \qquad h : S^1 \longrightarrow M$$

which defines a non-trivial cycle mod 2.

Claim: There exist in the space of unparametrized circles on S^1 three pairwise subordinated homology classes of dimensions (l-1, 2.(l-1), 3.(l-1) respectively which go under h into three pairwise subordinated homology classes of the space $\Lambda M/S^1 \times \theta$ of unparametrized closed curves on M. These give rise, according to (6.2), to at least three closed geodesics on M.

Note, however, that we did not exclude that two of these geodesics are multiple coverings of the same prime geodesic. This possibility seems highly unlikely, however. In the special case that M is a "pinched" manifold, one can exclude this:

6.4 THEOREM. Let $\pi_1 M = 0$, dim $M = n$, and assume that the sectional curvature K is > 0 and satisfies $\kappa \leq \min K : \max K$ with $\kappa \sim 0.64$. Then there exist on M g(n) prime closed geodesics of length in the interval $[2\pi, 4\pi[$, where $g(n) = 2n-s(n)-1$, $0 \leq s(n) < 2^h$.

We conclude our lecture with listing some recent results which imply the existence of an infinite number of prime closed geodesics on M.

6.5 THEOREM. Let $\pi_1 M < \infty$ and assume that the sequence of rational Betti numbers of ΛM is not bounded. Then there exist infinitely many prime closed geodesics on M.

This result of Gromoll and Meyer uses the Morse theory of E on ΛM. Although there are many examples of manifolds M satisfying the hypothesis of (6.5), there are also important cases where this is not satisfied, i.e., the sphere and the projective spaces. Recently, D. Sullivan has proved

6.6 LEMMA. *Let $\pi_1 M = 0$, M : compact. There exists a sequence $\{w_r\}$ of rational homology classes of ΛM where $\{\dim w_r\}$ forms an arithmetic sequence.*
Actually, the w_r can be represented by cycles lying in a subspace ΩM of ΛM.

This result has many interesting consequences:

6.7 THEOREM. *Assume that $\pi_1 M < \infty$ and that M has the rational homotopy type of a product of two simply connected finite CW-complexes. Then there are infinitely many prime closed geodesics on M.*

This follows using the theorem (6.5). An other consequence of (6.6) is

6.8 THEOREM. *Assume $\pi_1 M < \infty$. Then, generically, there are infinitely many prime closed geodesics on M.*

The proof uses the formula (4.10) for the index of the multiple coverings of a closed geodesic and the fact that the attaching maps in ΛM are equivariant with respect to the S^1- and θ-action on ΛM.

Using (6.6) one can also show.

6.9 LEMMA. *Assume* $\pi_1 M < \infty$. *Then there is no riemannian metric* g_o *on* M *such that* (M, g_o) *has only finitely many prime closed geodesics*, *all of which are hyperbolic*.

Using (5.6), this yields another proof of (6.8).

References:

[6.1] D. Gromoll - W. Meyer, Periodic geodesics on compact riemannian manifolds, J. Diff. Geometry 3 (1969), 493-510.

[6.2] P. Klein, Über die Kohomologie des freien Schleifenraums. Bonner Math. Schriften Nr. 55, Bonn 1972.

[6.3] W. Klingenberg, Simple closed geodesics on pinched spheres, J. Diff. Geom. 2 (1968), 225-232.

[6.4] W. Klingenberg, The space of closed curves on the sphere, Topology 7 (1968), 395-415.

[6.5] D. Sullivan, Differential forms and the topology of manifolds. Preprint M.I.T 1973.

W. Klingenberg

[6.6] A.S. Šcarc, Homology of the space of closed curves, Trudy Moskov. Mat. Obsc. 9 (1960), 3-44 (Russian).

CENTRO INTERNAZIONALE MATEMATICO ESTIVO

(C.I.M.E.)

VARIATIONAL METHODS FOR NONLINEAR EIGENVALUE PROBLEMS

P. H. RABINOWITZ

Corso tenuta a Varenna dal 16 al 25 giugno 1974

The goal of these lectures is to present an introduction to variational methods for nonlinear eigenvalue problems both in an abstract setting and as applied to nonlinear partial differential equations. Several different situations will be treated. Our study begins with "Theorems of Ljusternik-Schnirelmann type". The simplest such result, which is due to Ljusternik [1] states: If f is an even continuously differentiable real valued function on $\mathbb{R}^n$, then $f|_{S^{n-1}}$ possesses at least n distinct pairs of critical points (i.e. points at which $f'(x) = \lambda x$, $\lambda = (f'(x), x)$). This theorem serves as a prototype for more general situations where one has a real valued function (usually even) on a manifold (usually "spherelike") and uses topological invariants associated with the manifold to obtain lower bounds for the number of critical points the functional possesses. Morse theory treats similar questions and indeed there are many ideas in common. However smoothness requirements for theorems of Ljusternik-Schnirelmann type are less stringent (C^1 rather than C^2) than in Morse theory and critical points need not be nondegenerate. These facts combine to make the Ljusternik-Schnirelmann theory more applicable to nonlinear partial differential equations. On the other hand when it can be used, Morse theory gives more detailed information on critical points and their types (see e.g. [2]).

In addition to the usual type of Ljusternik-Schnirelmann theorems where the argument of the functional f is constrained to lie on a proper submanifold M of some underlying Banach or Hilbert space, E, thereby causing a Lagrange multiplier to appear where $f|_{M'} = 0$, we will also study the unconstrained case of a functional (usually even) defined on all of E. Qualitative information about the functional at 0 and ∞ can then be used to get lower bounds on the number of critical points the functional possesses. As a simple illustration suppose $f \in C^1(\mathbb{R}^n, \mathbb{R})$ with f even, $f(0) = 0$, $f > 0$ near 0, and $f < 0$ near ∞. Then aside from 0, f possesses at least n distinct pairs of critical points.

Often, in particular in applications to partial differential equations, f depends on a parameter and varying the parameter changes the topological structure of the level sets of f. This effect will also be studied.

The next set of topics to be treated concerns bifurcation. Suppose E is a real Hilbert space and $f \in C^1(E, \mathbb{R})$. Consider the equation

$$f'(u) = \lambda u \tag{0.1}$$

where $f'(u) = Lu + H(u)$, L being a (symmetric) linear operator and $H(u) = o(\|u\|)$ at $u = 0$. Therefore $u = 0$ and any $\lambda \in \mathbb{R}$ satisfies (0.1). We call this line of solutions the trivial solutions of (0.1). A point $(\mu, 0) \in \mathbb{R} \times E$ is said to be a bifurcation point if every neighborhood of $(\mu, 0)$ contains nontrivial solutions. A necessary condition that $(\mu, 0)$ be a bifurcation point is that μ belongs to the spectrum of L. Sufficient conditions have been obtained by various authors, e.g. [3], [4], [5]. In particular it has been shown by Böhme [6] and Marino [7] that every isolated eigenvalue μ of L of finite multiplicity corresponds to a bifurcation point. A proof of this result will be presented. In addition, we will briefly study the number of solutions of (0.1) as a function of λ near $(\mu, 0)$.

Lastly for a particular nonlinear partial differential equation, the interaction of variational and continuation methods will be investigated.

In §1 the technical preliminaries needed in the sequel will be presented. The notion of genus will be introduced and its properties developed. A deformation theorem which plays a crucial role in all later existence results will also be proved. The next two sections treat the theorems of Ljusternik-Schnirelmann type. The constrained case will be treated in §2 and the general Banach space case in §3. Applications of each of the general theorems to nonlinear partial differential equations will be given as the theory develops. Bifurcation results will be presented in §4. Lastly in §5 the connection between variational methods and positivity arguments for a particular class of partial differential equations will be explored.

For other general references on the above topics, we suggest the book of Ljusternik-Schnirelmann [8], Krasnoselski [3], Vainberg [9], Schwartz [10], and Fučik, Nečas, J. Souček and V. Souček [11] as well as the papers of Palais [2] [12], Browder [13] [14], Ambrosetti and Rabinowitz [15], and Berger [16].

1. Preliminary material.

In this section we introduce the notion of genus which provides the topological basis for many of the existence results to be presented later. A general deformation theorem used repeatedly in the sequel will also be proved.

First some notation. For X and Y Banach spaces and $U \subset X$, $V \subset Y$, $C^k(U,V)$ denotes the space of k times continuously Frechet differentiable maps from U to V. If $f \in C^k(U,V)$ and $u \in U$, $f'(u)$ denotes the Frechet derivative of f at u. If X' is the dual of X, $\langle \cdot,\cdot \rangle$ denotes the pairing between X' and X while if X is a Hilbert space, $(\cdot,\cdot)$ will denote the inner product in X. For $x \in X$ and $r > 0$, $B_r(x) = \{y \in X \mid \|x - y\| < r\}$ and $B_r \equiv B_r(0)$ with boundary S_r. The interior of a set U will be denoted by int U. Lastly for $K \subset X$ and $\delta > 0$, $N_\delta(K) = \{x \in X \mid \|x - K\| \leq \delta\}$ where $\|x - K\|$ denotes the distance from x to the set K.

Let E be a real Banach space and $\Sigma(E) \equiv \Sigma$ the collection of $A \subset E - \{0\}$ with A closed in E and symmetric with respect to the origin, i.e. $-x \in A$ whenever $x \in A$. The set $A \subset \Sigma$ is said to have <u>genus n</u> (denoted by $\gamma(A) = n$) if there exists an odd map $\varphi \in C(A, \mathbb{R}^n - \{0\})$ and n is the smallest integer having this property. If $A = \phi$, we write $\gamma(A) = 0$ and if there is no finite such n, we set $\gamma(A) = \infty$.

The definition of genus given here is essentially that used by Coffman [17]. (See also the coindex of Connor and Floyd [18].) It is equivalent to an earlier definition given by Krasnoselski [3], a proof of the equivalence being given in [19].

The following lemma gives properties of genus which will be needed later.

<u>Lemma 1.1</u>: Let $A, B \in \Sigma$.

1° If there is an odd $f \in C(A, B)$, then $\gamma(A) \leq \gamma(B)$.

2° If $A \subset B$, $\gamma(A) \leq \gamma(B)$.

3° If there is an odd homeomorphism $h \in C(A, B)$, then $\gamma(A) = \gamma(B)$.

4° $\gamma(A \cup B) \leq \gamma(A) + \gamma(B)$.

5° If $\gamma(B) < \infty$, $\gamma(\overline{A - B}) \geq \gamma(A) - \gamma(B)$.

6° If A is compact, $\gamma(A) < \infty$ and there exists a $\delta > 0$ such that $\gamma(N_\delta(A)) = \gamma(A)$.

7° If $\gamma(A) > k$, V is a k dimensional subspace of E, and $V^{\perp}$ is an algebraically and topologically complementary subspace, then $A \cap V^{\perp} \neq \phi$.

Proof: If $\gamma(A)$ or $\gamma(B) = \infty$, the results become trivial. Hence we assume for what follows that both $\gamma(A)$, $\gamma(B) < \infty$.

1° Suppose $\gamma(B) = n$. Then there exists $\varphi \in C(B, \mathbb{R}^n - \{0\})$ with φ odd. The map $\varphi \circ f \in C(A, \mathbb{R}^n - \{0\})$ and is odd. Hence $\gamma(A) \leq n = \gamma(B)$.

2° Take f to be the identity map in 1°.

3° Immediate from 1° on interchanging the roles of A and B.

4° Suppose $\gamma(A) = n$, $\gamma(B) = m$. Then there exists $\varphi \in C(A, \mathbb{R}^n - \{0\})$, $\psi \in C(B, \mathbb{R}^m - \{0\})$ with φ and ψ odd. Using the Tietze extension theorem extend φ, ψ to $\hat{\varphi} \in C(E, \mathbb{R}^n)$, $\hat{\psi} \in C(E, \mathbb{R}^m)$ respectively. Replacing $\hat{\varphi}(x)$ by $\frac{1}{2}(\hat{\varphi}(x) - \hat{\varphi}(-x))$ it can be assumed that $\hat{\varphi}$ and similarly $\hat{\psi}$ are odd. Then $f = (\hat{\varphi}, \hat{\psi}) \in C(A \cup B, \mathbb{R}^{n+m} - \{0\})$ with f odd. Hence $\gamma(A \cup B) \leq n + m = \gamma(A) + \gamma(B)$.

5° Since $A \subset \overline{A - B} \cup B$, the result follows from 2° and 4°.

6° If $x \in E$, $0 < r < \|x\|$, and $C = B_r(x) \cup B_r(-x)$, then $\overline{B_r(x)} \cap \overline{B_r(-x)} = \phi$ so $\gamma(\overline{C}) = 1$. Since A is compact, it can be covered by finitely many such C. Hence $\gamma(A) < \infty$ via 4°. If $\gamma(A) = n$, there is an odd $\varphi \in C(A, \mathbb{R}^n - \{0\})$. Extend φ oddly to $\hat{\varphi} \in C(E, \mathbb{R}^n)$ as in 4°. Since $\hat{\varphi} \neq 0$ on A, the compactness of A implies there exists $\delta > 0$ such that $\hat{\varphi} \neq 0$ on $N_\delta(A)$. Therefore $\gamma(N_\delta(A)) \leq n = \gamma(A)$. The reverse inequality follows from 2°, giving the result.

7° Let P denote the projector of E onto V along $V^{\perp}$. If $A \cap V^{\perp} = \phi$ then $P \in C(A, V - \{0\})$ with P odd. By 3°, $\gamma(A) \leq k$, a contradiction.

For later sections it is important to calculate the genus of some special sets. This is carried out next.

Theorem 1.2: Let Ω be bounded open symmetric $\subset \mathbb{R}^n$ with $0 \in \Omega$ and let $A \in \Sigma$ be homeomorphic to $\partial\Omega$ by an odd homeomorphism h. Then $\gamma(A) = n$.

Proof: Recall the following version of the Borsuk-Ulam Theorem [10]: If Ω is a bounded open symmetric neighborhood if 0 in $\mathbb{R}^m$ and $\psi \in C(\partial\Omega, \mathbb{R}^m)$ with ψ odd and $\psi(\partial\Omega)$ contained in a proper subspace of $\mathbb{R}^m$, then there is $x \in \partial\Omega$ such that $\psi(x) = 0$. To prove the theorem, observe that since $h \in C(A, \partial\Omega)$, $\gamma(A) \leq n$ by 1° of Lemma 1.1. If $\gamma(A) = j < n$ there is a $\varphi \in C(A, \mathbb{R}^j - \{0\})$ with φ odd. Hence $\psi = \varphi \circ h^{-1} \in C(\partial\Omega, \mathbb{R}^j - \{0\})$ violating the Borsuk-Ulam Theorem.

A useful special case of Theorem 1.2 is

Corollary 1.3: If $A \in \Sigma$ is homeomorphic to S^{n-1} by an odd homeomorphism, $\gamma(A) = n$.

The notion of genus is closely related to the notion of category of Ljusternik-Schnirelmann. See e.g. [19].

The remainder of this section is devoted to proving a deformation theorem. Let E be a real Banach space, $U \subset E$ and $\Phi \in C^1(U, \mathbb{R})$. The following notion is due to Palais, e.g. [12]. (See also Browder [14].):

$v \in E$ is called a pseudogradient vector for Φ at $u \in U$ if

$$(1.5) \qquad \begin{cases} \text{(i)} & \|v\| \leq 2\|\Phi'(u)\| \\ \text{(ii)} & \langle \Phi'(u), v \rangle \geq \|\Phi'(u)\|^2 . \end{cases}$$

We are not distinguishing in (1.5) or in the sequel between the norms in E and in E'. It will be clear from the context which is meant. Note that a pseudogradient (p.g.) vector is not unique in general and a convex combination of p.g. vectors for Φ at u is also a p.g. vector for Φ at u. If $\Phi \in C^1(E,\mathbb{R})$, $\tilde{E} = \{u \in E \mid \Phi'(u) \neq 0\}$, $v(x)$ is called a pseudogradient vector field on $\tilde{E}$ if v is locally Lipschitz continuous and $v(x)$ is a p.g. vector for all $x \in \tilde{E}$.

Lemma 1.6: If $\Phi \in C^1(E,\mathbb{R})$, there exists a pseudogradient vector field for Φ on $\tilde{E}$.

Proof: Let $u \in \tilde{E}$ and $w \in E$ with $\|w\| = 1$ and $\langle \Phi'(u), w \rangle > \frac{2}{3}\|\Phi'(u)\|$. Then

$$(1.7) \qquad z = \frac{3}{2}\|\Phi'(u)\| w$$

is a p.g. vector for Φ at u with strict inequality in (1.5). By the continuity of Φ', z is a p.g. vector for Φ for all v in an open neighborhood of N_u of u. The set of all such neighborhoods covers $\tilde{E}$. Therefore there exists a locally finite refinement $\{N_{u_i}\}$. Let $\rho_i(x)$ denote the distance from x to the complement of N_{u_i}. Then ρ_i is Lipschitz continuous and vanishes outside of N_{u_i}. Set

$$\beta_i(x) = \frac{\rho_i(x)}{\sum_r \rho_r(x)}$$

Since $\{N_{u_i}\}$ is a locally finite covering, for each $x \in \tilde{E}$, the denominator of $\beta_i(x)$ is only a finite sum. Finally let $v(x) = \sum_i z_i \beta_i(x)$ where z_i is given by (1.7) for $u = u_i$. For each $x \in \tilde{E}$, v is a convex combination of p.g. vectors for Φ and hence is a p.g. vector. Moreover v is locally Lipschitz continuous. The proof is complete.

Corollary 1.8: If Φ is even, v can be chosen to be odd.

Proof: If Φ is even, replace v by $w(x) = \frac{1}{2}(v(x) - v(-x))$. Then w is odd, locally Lipschitz continuous, and (1.5) is satisfied since

(i) $\quad \|w(x)\| \leq \frac{1}{2}(\|v(x)\| + \|v(-x)\|) \leq 2\|\Phi'(x)\|$

(ii) $\quad \langle \Phi'(x), w(x) \rangle = \frac{1}{2}[\langle \Phi'(x), v(x) \rangle + \langle \Phi'(-x), v(-x) \rangle] \geq \|\Phi'(x)\|^2$.

Next let $f \in C^1(E, \mathbb{R})$. Recall u is a critical point for f if $f'(u) = 0$ and c is a critical value for f if $f^{-1}(c)$ contains a critical point. Set $A_c = \{x \in E \mid f(x) \leq c\}$ and $K_c = \{x \in E \mid f(x) = c,\ f'(x) = 0\}$, i.e. K_c is the set of critical values corresponding to c. We say f satisfies the Palais-Smale condition (P.S.) if any sequence (u_n) along which $|f(u_n)|$ is bounded and $f'(u_n) \to 0$ possesses a convergent subsequence. If this condition is satisfied only in the region where $f \geq \alpha > 0$ (resp. $f \leq -\alpha < 0$) for all $\alpha > 0$, we say f satisfies $(PS)^+$ $((PS)^-)$.

Now we are ready for the deformation theorem. A slightly strengthened version of a result due to Clark [20] will be given. See [20] for references to earlier such results.

Theorem 1.9: Let $f \in C^1(E, \mathbb{R})$ and satisfy (P.S.). If $c \in \mathbb{R}$ and N is any neighborhood of K_c, there exists $\eta(t,x) \equiv \eta_t(x) \in C([0,1] \times E, E)$ and constants $\bar{\varepsilon} > \varepsilon > 0$ such that

1° $\eta_0(x) = x$ for all $x \in E$.

2° $\eta_t(x) = x$ for all $x \in f^{-1}[c-\bar{\varepsilon}, c+\bar{\varepsilon}]$ and for all $t \in [0,1]$.

3° $\eta_t(x)$ is a homeomorphism of E onto E for all $t \in [0,1]$.

4° $f(\eta_t(x)) \le f(x)$ for all $x \in E$, $t \in [0,1]$.

5° $\eta_1(A_{c+\varepsilon} - N) \subset A_{c-\varepsilon}$.

6° If $K_c = \phi$, $\eta_1(A_{c+\varepsilon}) \subset A_{c-\varepsilon}$.

7° If f is even, η_t is odd in x.

Proof: We assume $K_c \neq \phi$. The remaining case is simpler. By (P.S.), K_c is compact. Hence for $0 < \delta$ sufficiently small, $M_\delta \equiv \text{int}\, N_\delta(K_c) \subset N$ so it suffices to prove 5° with N replaced by M_δ.

There are constants $b, \bar{\varepsilon} > 0$ (and depending on δ) such that

$$(1.10) \qquad \|f'(x)\| \ge b \quad \text{for } x \in A_{c+\bar{\varepsilon}} - A_{c-\bar{\varepsilon}} - M_{\delta/8}$$

for otherwise there exist sequences $b_n \to 0$, $\varepsilon_n \to 0$, and $x_n \in A_{c+\varepsilon_n} - A_{c-\varepsilon_n} - M_{\delta/8}$ with $\|f'(x_n)\| < b_n$. By (P.S.), a subsequence of x_n converges to x satisfying $f(x) = c$, $f'(x) = 0$, and $x \notin M_{\delta/8}$. But $x \in K_c \subset M_{\delta/8}$, a contradiction. Since (1.10) remains valid if $\bar{\varepsilon}$ is decreased, we can assume

$$(1.11) \qquad 0 < \bar{\varepsilon} < \min\left(\frac{b\delta}{2}, \frac{b^2}{8}, \frac{1}{8}\right).$$

Let $\varepsilon \in (0, \bar{\varepsilon})$, $A = \{x \in E \mid f(x) \ge c + \bar{\varepsilon} \text{ or } f(x) \le c - \bar{\varepsilon}\}$ and $B = \{x \in E \mid c - \varepsilon \le f(x) \le c + \varepsilon\}$. Therefore $A \cap B = \phi$. Define $g(x) = \|x - A\| \left(\|x - A\| + \|x - B\|\right)^{-1}$. Then g is Lipschitz continuous with $g = 0$ on A, $g = 1$ on B, and $0 \le g(x) \le 1$. In a similar fashion there is a Lipschitz continuous $\bar{g}$ with $\bar{g} = 1$ on $E - M_{\delta/4}$, $\bar{g} = 0$ on

$M_{\delta/8}$, and $0 \le \overline{g}(x) \le 1$. Observe that if f is even, A, B, M_δ will be symmetric and $g, \overline{g}$ can be taken to be even functions. Next define $h(s) = 1$ if $s \in [0,1]$ and $h(s) = s^{-1}$ if $s \ge 1$ so h is Lipschitz continuous. Since $f \in C^1(E, \mathbb{R})$, by Lemma 1.6 there exists a p.g. vector field v for f on $\tilde{E}$ (odd if f is even). Finally set $V(x) = -g(x)\overline{g}(x)h(\|v(x)\|)v(x)$. Then V is a locally Lipschitz continuous vector field on E with $0 \le \|V(x)\| \le 1$ and odd if f is even.

Consider the ordinary differential equation

$$\text{(1.12)} \qquad \frac{d\eta}{dt} = V(\eta) , \quad \eta(0,x) = x \quad \text{for } x \in E .$$

By the basic existence theorem for such equations, for each $x \in E$ there exists a unique solution $\eta(t,x)$ of (1.12) for $t \in (t^-(x), t^+(x))$, a maximal t interval depending on x. Since V is bounded, $t^{\pm}(x) = \pm\infty$ for all $x \in E$ for otherwise if e.g. $t^+(x) < \infty$, let $t_n \to t^+(x)$. Integrating (1.12),

$$\text{(1.13)} \qquad \|\eta(t_{n+1}, x) - \eta(t_n, x)\| = \|\int_{t_n}^{t_{n+1}} V(\eta(t,x))\,dt\| \le |t_{n+1} - t_n| .$$

This implies $\eta(t_n, x)$ is a Cauchy sequence converging to $\overline{x}$ as $t_n \to t^+(x)$. But then the solution of (1.12) with initial condition $\overline{x}$ furnishes a continuation of $\eta(t,x)$ contradicting the maximality of $t^+(x)$.

It now follows in particular that $\eta_t(x) \in C([0,1] \times E, E)$ and satisfies $1°$. Since $g = 0$ on A, $2°$ obtains. The semigroup property for solutions of ordinary differential equations implies $3°$ and above remarks yield $7°$. To verify $4°$, note first that it is trivial when $\eta_t(x) \equiv x$. If $\eta_t(x) \not\equiv x$, $v(\eta_t(x))$ is defined and

$$(1.14)\qquad \frac{d}{dt} f(\eta_t(x)) = \langle f'(\eta_t(x)), \frac{d}{dt}\eta_t(x)\rangle$$

$$= \langle f'(\eta_t(x)), -g(\eta_t(x))\overline{g}(\eta_t(x))h(\|v(\eta_t(x))\|)v(\eta_t(x))\rangle$$

$$= -g(\eta_t(x))\overline{g}(\eta_t(x))h(\|v(\eta_t(x))\|)\langle f'(\eta_t(x)), v(\eta_t(x))\rangle \le 0$$

(actually < 0) by the definitions of $g, \overline{g}, h,$ and (1.5).

Lastly we must show

$$(1.15)\qquad \eta_1(A_{c+\varepsilon} - M_\delta) \subset A_{c-\varepsilon}\ .$$

Since by 4°, $f(\eta_t(x)) \le c - \varepsilon$ for $x \in A_{c-\varepsilon}$, (1.15) need only be verified for $x \in Y \equiv A_{c+\varepsilon} - A_{c-\varepsilon} - M_\delta$. Setting $\alpha_x(t) = f(\eta_t(x))$, this amounts to showing $\alpha_x(1) \le c - \varepsilon$ for $x \in Y$. By (1.14), $\frac{d}{dt}\alpha_x(t) \le 0$. Since $g \equiv 0$ on $A_{c-\overline{\varepsilon}}$, the orbit $\eta_t(x)$ cannot enter $A_{c-\varepsilon}$ so on integration this yields

$$(1.16)\qquad 0 \le \alpha_x(0) - \alpha_x(t) \le 2\overline{\varepsilon}\ .$$

If $x \in Y$ and $\eta_s(x) \in Z \equiv A_{c+\varepsilon} - A_{c-\varepsilon} - M_{\delta/2}$ for $s \in [0,t]$ (which is certainly the case for small t) then by (1.10), $\eta_s(x) \in \tilde{E}$ and $g(\eta_s(x)) = \overline{g}(\eta_s(x)) = 1$ for $s \in [0,t]$. Therefore

$$(1.17)\qquad 2\overline{\varepsilon} \ge \int_0^t h(\|v(\eta_s(x))\|)\langle f'(\eta_s(x)), v(\eta_s(x))\rangle ds$$

$$\ge \int_0^t h(\|v(\eta_s(x))\|)\|f'(\eta_s(x))\|^2 \qquad \text{(by (1.5))}$$

$$\ge b\int_0^t h(\|v(\eta_s(x))\|)\|f'(\eta_s(x))\|ds \qquad \text{(by (1.10))}$$

$$\ge \frac{b}{2}\int_0^t h(\|v(\eta_s(x))\|)\|v(\eta_s(x))\|ds \qquad \text{(by (1.5))}$$

$$\geq \frac{b}{2} \left\| \int_0^t h(\|v(\eta_s(x))\|) v(\eta_s(x)) ds \right\| = \frac{b}{2} \left\| \int_0^t V(\eta_s(x)) \right\|$$

$$= \frac{b}{2} \|\eta_t(x) - x\| .$$

Combining (1.17) and (1.11) yields

$$\|\eta_t(x) - x\| < \frac{\delta}{8} . \tag{1.18}$$

In particular the orbit $\eta_t(x)$ cannot enter $M_{\delta/2}$ and therefore can only leave Z by entering $A_{c-\varepsilon}$. To show that this in fact occurs for $t \in (0,1)$, suppose $\eta_t(x) \in Z$ for all $t \in [0,1]$. From (1.14) again

$$\frac{d\alpha_x(t)}{dt} \leq -h(\|v(\eta_t(x))\|) \|f'(\eta_t(x))\|^2 . \tag{1.19}$$

If for some t, $\|v(\eta_t(x))\| \leq 1$, $h(\|v(\eta_t(x))\|) = 1$ and by (1.19) and (1.10),

$$\frac{d\alpha_x(t)}{dt} \leq -\|f'(\eta_t(x))\|^2 \leq -b^2 \tag{1.20}$$

while if $\|v(\eta_t(x))\| > 1$, $h(\|v(\eta_t(x))\|) = \|v(\eta_t(x))\|^{-1}$ and by (1.19) and (1.5),

$$\frac{d\alpha_x(t)}{dt} \leq -\frac{\|f'(\eta_t(x))\|^2}{\|v(\eta_t(x))\|} \leq -\frac{\|v(\eta_t(x)\|}{4} \leq -\frac{1}{4} . \tag{1.21}$$

Combining (1.20) and (1.21) produces

$$\frac{d\alpha_x}{dt} \leq -\min(b^2, \tfrac{1}{4}) \tag{1.22}$$

so

$$\min(b^2, \tfrac{1}{4}) \leq \alpha_x(0) - \alpha_x(1) \leq 2\bar{\varepsilon} \tag{1.23}$$

which violates (1.11). The proof is complete.

Remark 1.24: (i) If $c > 0$ $(c < 0)$, (PS) can be replaced by $(PS)^+$ $((PS)^-)$ and $\bar{\varepsilon}$ can be chosen in $(0,c)$ $((c,0))$. This will be the case for some of our later applications. Actually (PS) is only needed for $f^{-1}([c-\bar{\varepsilon}, c+\bar{\varepsilon}])$.

(ii) If V is replaced by $-V$, Theorem 1.9 is valid with η replaced by $\tilde{\eta}$, a reversal of the inequalities in 4°, ε replaced by $-\varepsilon$ in 5°, 6°, and A_c replaced by $\tilde{A}_c = \{x \in E \mid f(x) \geq c\}$.

(iii) Although Theorem 1.9 is set in the framework of a real Banach space, the ideas contained there can be applied more widely. For example suppose E is a real Hilbert space and $S_r = \partial B_r$. Let $\tilde{f} = f|_{S_r}$. Then $\tilde{f}'(x) = f'(x) - r^{-2}(f'(x), x)x$. Set $\tilde{K}_c = \{x \in S_r \mid f(x) = c,$ $f'(x) = r^{-2}(f'(x), x)x\}$. Then provided that $\tilde{f}$ satisfies (PS) and $A_{c+\varepsilon}$, N, V, etc. are replaced by their appropriate relativizations to S_r, there is an $\tilde{\eta} \in C([0,1] \times S_r, S_r)$ having the properties stated in Theorem 1.9 and only minor modifications are required for the proof. Use of this result will be made in the next section.

2. Theorems of Ljusternik-Schnirelmann type - the constrained case.

In this section we begin our study of theorems of Ljusternick-Schnirelmann type. The case of even functionals on a sphere, historically the first case treated, will be presented here as well as applications to nonlinear elliptic partial differential equations.

To display the ideas in the simplest setting, suppose $f \in C^1(\mathbb{R}^n, \mathbb{R})$ with f even. At a critical point of $\tilde{f} = f|_{S_r}$ we have

$$f'(x) = \lambda x\,, \quad \lambda = r^{-2}(f'(x), x)\,. \tag{2.1}$$

How many critical points does $\tilde{f}$ possess? Clearly $\tilde{f}$ has a maximum and a minimum and since f is odd, the corresponding critical points occur in antipodal pairs. If $f(x) = \frac{1}{2}(Lx, x)$ with L an $n \times n$ symmetric matrix, (2.1) becomes $Lx = \lambda x$ so x is an eigenvector and λ an eigenvalue of L. Hence for this case $\tilde{f}$ possesses at least n distinct pairs of critical points. Surprisingly this turns out to be a general fact.

<u>Theorem 2.2</u>: Let $f \in C^1(\mathbb{R}^n, \mathbb{R})$ with f even. Then for each $r > 0$, $\tilde{f}$ possesses at least n distinct pairs of critical points.

Proof: For convenience we take $r = 1$ so $S_1 = S^{n-1}$. The proof consists of three steps. First a minimax characterization of n "critical values", $b_1, \ldots, b_n$, of $\tilde{f}$ is given and their properties are studied. Secondly, it is shown that $b_1, \ldots, b_n$ are indeed critical values of $\tilde{f}$. Lastly for the degenerate case in which not all $b_1, \ldots, b_n$ are distinct the existence of infinitely many critical points of f is proved. This last step corresponds in the quadratic case to the fact that a multiple eigenvalue of L has an eigenspace of dimension greater than one.

(a) Minimax characterization of critical values of $\tilde{f}$.

Let $\Upsilon_k = \{A \subset S^{n-1} \mid \gamma(A) \geq k\}$. By Corollary 1.3, $\Upsilon_k \neq \phi$, $1 \leq k \leq n$. Define

$$b_k = \inf_{A \in \Upsilon_k} \max_{x \in A} f(x) \qquad 1 \leq k \leq n. \tag{2.3}$$

We will show that b_k are critical values of $\tilde{f}$. First some observations about these numbers. If $\gamma(A) > k$, it is not difficult to show there exists $B \subset A$ such that $\gamma(B) = k$. Hence $\max_B f \leq \max_A f$ so b_k can also be characterized as

(2.4) $$b_k = \inf_{B\in\gamma_k,\ \gamma(B)=k} \max_{x\in B} f(x)$$

Since b_{k+1} is an infimum over a smaller class of sets than b_k, $b_1 \le b_2 \le \dots \le b_n$. Next note that for any $x \in S^{n-1}$, $A = \{x,-x\} \in \gamma_1$. Hence $b_1 = \min_{S^{n-1}} f(x)$. Moreover $b_n = \max_{S^{n-1}} f(x)$. To verify this, it suffices to show that S^{n-1} is the only set in γ_n. Suppose $A \subset S^{n-1}$, inclusion being proper. The coordinate system can be assumed oriented so that $(0,\dots,0,\pm 1) \notin A$. Define $p(x_1,\dots,x_n) = (x_1,\dots,x_{n-1},0)$. Therefore $p \in C(A,\mathbb{R}^{n-1} - \{0\})$ and is odd so $\gamma(A) \le n-1$.

There is an equivalent characterization of the b_k that will be used later. Set $\tilde{A}_c = \{x \in S^{n-1} \mid f(x) \le c\}$. Then

(2.5) $$b_k = \inf\{r \in \mathbb{R} \mid \gamma(\tilde{A}_r) \ge k\} .$$

Indeed let r_k denotes the right hand side of (2.5). Suppose $\gamma(\tilde{A}_r) \ge k$. Then $\tilde{A}_r \in \gamma_k$ and $b_k \le \max_{\tilde{A}_r} f = r$. Since this is valid for all such r, $b_k \le r_k$. If $b_k < r_k$, there is an $A \in \gamma_k$ such that $\max_A f = s < r_k$. Hence $A \subset \tilde{A}_s$ and by 2° of Lemma 1.1, $\gamma(\tilde{A}_s) \ge \gamma(A) \ge k$. But then $r_k \le s$, a contradiction.

Another set of critical values of f is given by

(2.6) $$c_j = \sup_{Q\in\gamma_j} \min_{x\in Q} f(x) \qquad 1 \le j \le k .$$

This follows from the b_j case on replacing f by $-f$. It is easy to see that $c_n = b_1 \le \dots \le c_1 = b_n$. However simple examples show in general $c_j \neq b_{n+1-j}$ for $j \neq 1, n$.

(b) b_k is a critical value of $\tilde{f}$.

Since S^{n-1} is compact is is trivial to verify that $\tilde{f}$ satisfies (P.S.) so by Remark 1.24(iii), Theorem 1.9 is applicable here. If b_k is not a critical value of $\tilde{f}$, by 6° of Theorem 1.9, there is an $\varepsilon > 0$ and $\eta_1 \in C(E,E)$ such that $A = \eta_1(\tilde{A}_{b_k+\varepsilon}) \subset \tilde{A}_{b_k-\varepsilon}$. Since (2.5) shows $\gamma(\tilde{A}_{b_k+\varepsilon}) \geq k$ and η_1 is odd, by 1° of Lemma 1.1, $\gamma(A) \geq k$. Hence $A \in \gamma_k$ but

$$b_k \leq \max_A f \leq \max_{\tilde{A}_{b_k-\varepsilon}} f \leq b_k - \varepsilon$$

a contradiction. Thus b_k is a critical value of $\tilde{f}$.

(c) A multiplicity lemma.

It may happen that b_k is a multiple critical value of $\tilde{f}$, i.e. $b_k = \ldots = b_{k+p-1} \equiv b$ with perhaps only one pair of corresponding critical points. Then (b) is not sufficient to prove the assertion of Theorem 2.2. The following lemma completes the proof.

<u>Lemma 2.7</u>: If $b_k = \ldots = b_{k+p-1} \equiv b$ and $\tilde{K}_b = \{x \in S^{n-1} \mid \tilde{f}(x) = b, \tilde{f}'(x) = 0\}$ then $\gamma(\tilde{K}_b) \geq p$.

Proof: If $\gamma(\tilde{K}_b) \leq p - 1$, by 6° of Lemma 1.1, there is a $\delta > 0$ such that $\gamma(N_\delta(\tilde{K}_b)) \leq p - 1$. By 5° of Theorem 1.9 with $N = \text{int}\, N_\delta(\tilde{K}_b)$, there exists an $\varepsilon > 0$ such that $\eta_1(\tilde{A}_{b+\varepsilon} - N) \equiv Q \subset \tilde{A}_{b-\varepsilon}$. Since $b = b_{k+p-1}$, (2.5) shows that $\gamma(\tilde{A}_{b+\varepsilon}) \geq k + p - 1$. Therefore by 5° and 1° of Lemma 1.1, $\gamma(\tilde{A}_{b+\varepsilon} - N) = \gamma(\overline{A_{b+\varepsilon} - N}) \geq (k + p - 1) - (p - 1) = k$ and $\gamma(Q) \geq k$. Hence $Q \in \gamma_k$ and

$$b_k \leq \max_Q f \leq \max_{\tilde{A}_{b-\varepsilon}} \leq b_k - \varepsilon$$

a contradiction.

Remark 2.8: (i) If $p > 1$ in Lemma 2.7, $\tilde{K}_b$ contains infinitely many critical points since a finite set has genus 1.
(ii) If S^{n-1} is replaced by a more general C^1 manifold given by $g(x) = r$ which is diffeomorphic to S^{n-1} and $(g'(x),x) > 0$ for $x \neq 0$, the above proof goes through virtually unchanged where now $f|'_{\{g(x)=r\}} = f' - \lambda g'$, $\lambda = (f'(x),x)\,(g'(x),x)^{-1}$.

For the infinite dimensional analogue of Theorem 2.2, more structure is needed. If E is a real Hilbert space and $f \in C^1(E,\mathbb{R})$, then for each $u \in E$, $f'(u)$ is a linear map from E to $\mathbb{R}$, i.e. $f'(u) \in E'$. Since E is self dual, $f'(u)$ can be identified with an element of E. However it should be kept in mind that properly speaking (2.1) means

$$(2.9) \qquad (f'(u),e) = \lambda(u,e) \quad \text{for all } e \in E .$$

Possible critical values of f can again be characterized by (2.3) with max replaced by sup or by (2.5) (as well as (2.6) with min replaced by inf). An examination of the proofs of (b), (c) of Theorem 2.2 then shows we get the following abstract result.

Theorem 2.10: Suppose E is a real infinite dimensional Hilbert space, $f \in C^1(E,\mathbb{R})$ is even, $f|_{S_r}$ satisfies (PS) and is bounded from below. Then $f|_{S_r}$ possesses infinitely many distinct pairs of critical points.

For Theorem 2.10 to be useful, conditions on f are required under which the above hypotheses are satisfied. Lemma 2.11 gives one such set essentially due to Krasnoselski [3].

Lemma 2.11: Let E be a real infinite dimensional separable Hilbert space, $f \in C^1(E,\mathbb{R})$ be even with $f(0) = 0$ and $f(x) < 0$ if $x \neq 0$. Suppose further that f' maps weakly convergent to strongly convergent sequences and $f'(x) \neq 0$ if $x \neq 0$. Then f satisfies the hypotheses of Theorem 2.10 for all $r > 0$.

Proof: Again for convenience we take $r = 1$ and $S_1 \equiv S$. Since f' maps weakly convergent to strongly convergent sequences, f' is compact. A standard result [9] implies f is weakly continuous. If $\inf_S f = -\infty$, there is a sequence $(u_n) \subset S$ such that $f(u_n) < -n$. Since (u_n) is bounded and E is a Hilbert space, (u_n) possesses a weakly convergent subsequence $u_{n_i} \rightharpoonup u$. Therefore $f(u_{n_i}) \to f(u) = -\infty$ which is not possible. Hence f is bounded from below.

Since $f < 0$ on S, to complete the proof it suffices to verify $(PS)^-$. Thus suppose there is a sequence (u_n) and constants $K, \alpha > 0$ such that $-K \leq f(u_n) \leq -\alpha < 0$ and

$$f|_S'(u_n) = f'(u_n) - (f'(u_n), u_n)u_n \to 0 . \tag{2.12}$$

As above (u_n) possesses a subsequence $u_{n_i} \rightharpoonup u$ and $f(u_{n_i}) \to f(u) \leq -\alpha$. Therefore $u \neq 0$ and $f'(u) \neq 0$. Moreover $f'(u_{n_i}) \to f'(u)$ and $(f'(u_{n_i}), u_{n_i}) \to (f'(u), u)$. Thus taking the inner product of (2.12) with $f'(u_n)$ and $n = n_i$ shows

$$0 \neq \|f'(u)\|^2 = (f'(u), u)^2 . \tag{2.13}$$

Thus $(f'(u_{n_i}), u_{n_i}) \neq 0$ for large n_i and dividing by this quantity in (2.12) yields the convergence of u_{n_i}. Hence $(PS)^-$.

<u>Remark 2.14</u>: (i) No explicit use of the separability of E was made above. Moreover as was shown by Krasnoselski [3], if E is not separable, there exists no f satisfying the above hypotheses.

(ii) Other conditions can be used to satisfy the hypotheses of Theorem 2.10. E.g. Amann [21] assumes f' maps weakly to strongly convergent sequences, $f(u) \neq 0$ implies $f'(u) \neq 0$, and $\gamma(\{x \in S \mid f(x) \neq 0\}) = \infty$. An examination of the proof of Lemma 2.11 shows it carries over to this case.

(iii) $(PS)^-$ implies each critical value b_k is of "finite multiplicity", i.e. $\gamma(K_{b_k}) < \infty$. Moreover $b_k \to 0$ as $k \to \infty$ as was shown by Krasnoselski [3]. This also follows as in the proof of Theorem 3.

(iv) Another set of critical values for $f|_S$ under the hypotheses of Lemma 2.11 can be obtained by defining

$$\tilde{b}_k = \inf_{A \in \tilde{\gamma}_k} \max_{u \in A} f(u) \tag{2.15}$$

where $\tilde{\gamma}_k = \{A \in S \mid A \text{ is compact and } \gamma(A) \geq k\}$. Clearly $\tilde{b}_k \geq b_k$ but we do not know if the two sets of critical values coincide. The $\tilde{b}_k$ can also be obtained by a Galerkin argument starting with Theorem 2.2 and passing to a limit. See [19].

(v) Within the context of Theorem 2.10 and Lemma 2.11, Fučik, Nečas, J. Souček, and V. Souček [22] have shown that under additional hypotheses, in particular real analyticity, $f|_S$ has at most countably many critical values.

(vi) For other more general abstract results, see e.g. [23], [2], [14].

Next Theorem 2.10 will be applied to nonlinear elliptic partial differential equations. For technical convenience here and in the future examples we stay within the framework of second order equations. Let $\Omega \subset \mathbb{R}^n$ be a bounded domain with smooth boundary. Consider

$$(2.15) \qquad Lu \equiv -\sum_{i,j=1}^{n} (a_{ij}(x)u_{x_j})_{x_i} + c(x)u = \lambda g(x,u) \qquad x \in \Omega$$

$$u = 0, \quad x \in \partial\Omega$$

where the functions $a_{ij}(x)$ and $c(x)$ are smooth, $c(x) \geq 0$ in $\overline{\Omega}$, and L is uniformly elliptic in $\overline{\Omega}$. These assumptions will also be made for all later applications. Concerning g, we assume

(g_1) $g(x,z)$ is locally Hölder continuous in $\overline{\Omega} \times \mathbb{R}$.

(g_2) $|g(x,z)| \leq \alpha_1 + \alpha_2 |z|^s$ where $1 \leq s < \frac{n+2}{n-2}$ for $n > 2$
$\leq \alpha_3 \exp \varphi(z)$ where $\varphi(z) z^{-2} \to 0$ as $z \to \infty$ for $n = 2$.

(g_3) g is odd in z.

(g_4) $zg(x,z) \geq 0$ and $= 0$ only at $z = 0$.

In (g_2) and the sequel, α_1, α_2, etc. repeatedly denote positive constants.

By a solution of (2.15) we mean a pair $(\lambda, u) \in \mathbb{R} \times C^2(\Omega)$. Let $E = W_0^{1,2}(\Omega)$. As norm in E we can take

$$\|u\|^2 = \int_\Omega \Big(\sum_{i,j=1}^{n} a_{ij}(x) u_{x_i} u_{x_j} + c(x) u^2 \Big) dx .$$

Let $G(x,z) = \int_0^z g(x,t)\,dt$. Then G is even in z and by (g_2)

$$|G(x,z)| \le \alpha_4 + \alpha_5 |z|^{s+1} \qquad \text{(for } n > 2) . \tag{2.16}$$

Let $f(u) = -\int_\Omega G(x,u(x))dx$. We will show f satisfies the hypotheses of Lemma 2.11 and this in turn leads to a solution of (2.15). For convenience in what follows we take $n > 2$. The $n = 2$ case follows similarly using the stronger version of Sobolev inequality for $n = 2$.

Formally,

$$\langle f'(u), \varphi \rangle = -\int_\Omega g(x,u(x))\varphi(x)dx . \tag{2.17}$$

By the Sobolev inequalities [24], $E \subset L^t(\Omega)$ for all $t \in [1, \frac{2n}{n-2}]$, the inclusion being continuous, i.e.

$$\left(\int_\Omega |\varphi|^t dx\right)^{1/t} \le c_t \|\varphi\| \tag{2.18}$$

for all $\varphi \in E$. Moreover for $t < \frac{2n}{n-2}$, the inclusion is compact. The growth conditions satisfied by g and G together with (2.18) and the Lebesgue dominated convergence theorem readily imply f is defined on E and $f'(u)$ exists for all $u \in E$. (g_3) implies f is even and (g_4) shows $f \le 0$ on E, $f(u) = 0$ only at $u = 0$, and $f'(u) = 0$ only at $u = 0$. That $f \in C^1(E, \mathbb{R})$ is a consequence of the following lemma which also shows f satisfies (PS).

<u>Lemma 2.19</u>: If g satisfies (g_1) and (g_2), f' maps weakly to strongly convergent sequences in E.

Proof:

$$\|f'(u)\| = \sup_{0 \neq \varphi \in E} \frac{\left|\int_\Omega g(x,u)\varphi \, dx\right|}{\|\varphi\|}$$

so by the Hölder inequality and (2.18),

$$\|f'(u)\| \leq \alpha_1 \left(\int_\Omega |g(x,u)|^{\frac{2n}{n+2}} dx\right)^{\frac{n+2}{2n}} . \tag{2.20}$$

Let $u_m \rightharpoonup u$. To complete the proof it suffices via (2.20) to show that

$$\int_\Omega |g(x,u_m) - g(x,u)|^{\frac{2n}{n+2}} dx \to 0 \tag{2.21}$$

as $m \to \infty$. By a theorem of Krasnoselski [3] if e.g. $q(x,z)$ is continuous on $\overline{\Omega} \times \mathbb{R}$ and

$$|q(x,z)| \leq \alpha_1 + \alpha_2 |z|^{\gamma/\delta}, \quad 1 \leq \gamma, \delta < \infty \tag{2.22}$$

then the map $w(x) \to q(x, w(x))$ is continuous from $L^\gamma(\Omega)$ to $L^\delta(\Omega)$. For $q = g$ in (2.22), $\delta = 2n(n+2)^{-1}$, and $\gamma = \delta s$ (which is permissible by (g_2)) we see (2.21) will be verified if $u_m \to u$ in $L^\gamma(\Omega)$. The compactness of the embedding of E in $L^t(\Omega)$ mentioned above shows $u_m \to u$ in $L^t(\Omega)$ for $t \in [1, \frac{2n}{n-2})$ and since $\delta s < \frac{2n}{n-2}$, the proof is complete.

The hypotheses of Theorem 2.10 being satisfied, for all $r > 0$, $f|_{S_r}$ possesses infinitely many distinct critical points. At these points, $f'(u) = \mu u$, i.e.

$$-\int_\Omega g(x,u)\varphi \, dx = \mu(u,\varphi) = \mu \int_\Omega \left(\sum_{i,j=1}^n a_{ij} u_{x_j} \varphi_{x_i} + cu\varphi \right) dx \tag{2.23}$$

for all $\varphi \in E$ where $\mu = -r^{-2} \int_\Omega g(x,u)u \, dx$. Setting $\lambda = \mu^{-1}$ we see (λ, u) is a weak solution of (2.15). Standard regularity arguments employing (g_1) and (g_2) then show $(\lambda, u) \in \mathbb{R} \times C^2(\Omega)$. Thus we have shown

Theorem 2.24: If g satisfies (g_1) - (g_4), then for each $r > 0$, (2.15) possesses infinitely many distinct pairs of solutions $(\lambda_m(r), \pm u_m(r))$ with $\| u_m(r) \| = r$.

Remark 2.25: For more general results of this nature, see e.g. Browder [13], [14].

3. Theorems of Ljusternik-Schnirelmann type - the free case.

In this section we continue our study of theorems of Ljusternik-Schnirelmann type. Let E be a real Banach space and $f \in C^1(E, \mathbb{R})$. The question of interest now is the existence of critical points of f in E; unlike §2 no constraints are imposed on f. The case where f is even will be our major concern, however some results for the general case will also be given.

The main difficulty in searching for critical points of f is in finding appropriate classes of sets to which the minimax arguments of §2 can be applied. There is one case however first studied by Clark [20] which can be treated as in §2 and we present it first. Let $\overline{\gamma}_k = \{A \subset E \mid \gamma(A) \geq k\}$.

Theorem 3.1: Let $f \in C^1(E, \mathbb{R})$ with f even and $f(0) = 0$. If f satisfies $(PS)^-$ and

$$-\infty < b_k \equiv \inf_{A \in \overline{\gamma}_k} \sup_{x \in A} f(x) < 0 ,$$

then b_k is a critical value of f. Moreover if $-\infty < b_k = \ldots = b_{k+p-1} \equiv b$, $\gamma(K_b) \geq p$.

Proof: It suffices to prove the second statement. Since this is identical to the proof of Theorem 2.2, we omit the details.

Next we will show how Theorem 3.1 can be applied to problems similar to (2.15). Consider

$$(3.2) \qquad Lu = \lambda(a(x)u + p(x,u)) \ , \quad x \in \Omega$$
$$u = 0 \ , \quad x \in \partial\Omega$$

where $a(x)$ is smooth and positive in $\overline{\Omega}$ and $p(x,z)$ satisfies

(p_1) $p(x,z)$ is locally Hölder continuous in $\Omega \times \mathbb{R}$ and $p(x,0) = 0$.

(p_2) $p(x,z) = o(|z|)$ at $z = 0$.

(p_3) p is odd in z.

(p_4) There exists a $\overline{z} > 0$ such that $a(x)\overline{z} + p(x,\overline{z}) < 0$.

Observe that (p_4) is satisfied if e.g. $p(x,z)z^{-1} \to -\infty$ as $z \to \infty$ (uniformly in $x \in \overline{\Omega}$).

Associated with (3.2) is the linear Sturm-Liouville eigenvalue problem:

$$(3.3) \qquad Lv = \lambda av \qquad x \in \Omega$$
$$v = 0 \qquad x \in \partial\Omega \ .$$

As is well known, (3.3) possesses a sequence of eigenvalues (λ_m) of finite multiplicity with $0 < \lambda_1 \le \dots \le \lambda_m \le \dots$ and $\lambda_m \to \infty$ as $m \to \infty$.

<u>Theorem 3.4</u>: If p satisfies (p_1) - (p_4) and $\lambda \in (\lambda_k, \lambda_{k+1}]$, then (3.2) possesses at least k distinct pairs of solutions.

Proof: We will put the problem into the framework of Theorem 3.1. Set $g(x,z) = a(x)z + p(x,z)$ for $|z| \le \overline{z}$, $g(x,z) = a(x)\overline{z} + p(x,\overline{z})$ for $z > \overline{z}$ and let g be odd in z. Consider (2.15) with this choice of g. If (λ, u)

is a solution with $\lambda > 0$, then $\max_{\Omega}|u(x)| < \bar{z}$. To see this, suppose the maximum of u occurs at $\bar{x}$ with $u(\bar{x}) > 0$. Rotating the coordinate system if necessary we can assume (as in the proof of the maximum principle) that L contains no mixed second particles so $0 \leq Lu(\bar{x})$. But then from (2.15) and (p_4) if $u(\bar{x}) \geq \bar{z}$,

$$Lu(\bar{x}) = \lambda p(\bar{x}, u(\bar{x})) < 0 ,$$

a contradiction. Thus to solve (3.2), it suffices to treat (2.15) with a bounded nonlinear term.

Let E and $G(x,z)$ be as in Theorem 2.24 and

$$f(u) = \frac{1}{2}\|u\|^2 - \lambda \int_{\Omega} G(x, u(x))\,dx . \tag{3.5}$$

Then $f(0) = 0$ and f is even. Since $g(x,z)$ is bounded,

$$|G(x,z)| \leq \alpha_1 + \alpha_2 |z| . \tag{3.6}$$

so as in §2, $f \in C^1(E, \mathbb{R})$. Note that

$$\langle f'(u), \varphi \rangle = (u, \varphi) - \lambda \int_{\Omega} p(x,u)\varphi\,dx \equiv (u - \lambda T(u), \varphi)$$

where T is compact. Let v_j denote a solution of (3.3) corresponding to λ_j and normalized so that $\|v_j\| = 1$. Suppose $v = \sum_{1}^{i} \beta_j v_j$ with $\max_{\Omega}|v(x)| < \bar{z}$. Then with P the primative of p,

$$(3.7)\qquad \begin{aligned} f(v) &= \frac{1}{2}\|v\|^2 - \lambda \int_\Omega (\frac{1}{2} av^2 + P(x,v))dx \\ &= \frac{1}{2}\|v\|^2 - \lambda \int_\Omega (\frac{1}{2} av^2 + o(|v(x)|^2)dx \\ &= \frac{1}{2}\sum_1^i (1 - \frac{\lambda}{\lambda_j})\beta_j^2 + o(\|v\|^2) \qquad \text{at } v = 0 \end{aligned}$$

where we have used the Poincare inequality. Hence $f(v) < 0$ and $b_i < 0$, $1 \leq i \leq k$.

It remains to verify $(PS)^-$. Suppose $f(u_m)$ is bounded. Then (3.6) and the form of f imply (u_m) is bounded. The compactness of T implies $T(u_m)$ possesses a convergent subsequence. Hence $f'(u_m) = u_m - \lambda T(u_m) \to 0$ and the above shows (u_m) possesses a convergent subsequence. Thus the hypotheses of Theorem 3.1 are satisfied and f possesses at least k distinct pairs of critical points. As in Theorem 2.24 they correspond to classical solutions of (3.2).

Remark 3.8: (i) Slightly different versions of Theorem 3.4 were given by Hempel [25] and Ambrosetti [26] using different arguments. The above theorem is taken from [28].

(ii) It is not difficult to show using variational or other methods that (3.2) possesses a "positive" (and "negative") solution for all $\lambda > \lambda_1$, i.e. a solution (λ, u) with $u > 0$ $(u < 0)$ in Ω. This is true even without (p_3) [19].

Next we treat a fairly simple and geometrical situation where it is not necessary for f to be even. Again suppose $f \in C^1(E, \mathbb{R})$ and

(f_1) $f(0) = 0$ and there exists $\rho, \alpha > 0$ with $f > 0$ in $B_\rho - \{0\}$, $f \geq \alpha$ on ∂B_ρ.

(f_2) There is an $e \in E$, $e \neq 0$, such that $f(e) = 0$.

Let $\Gamma = \{g \in C([0,1],E) \mid g(0) = 0,\ g(1) = e\}$ and define

$$b = \inf_{g \in \Gamma} \max_{u \in g([0,1])} f(u) .$$

Theorem 3.9: If $f \in C^1(E,\mathbb{R})$ and satisfies (f_1), (f_2), and $(PS)^+$, then b is a critical value of f.

Proof: (f_1) implies $b \geq \alpha > 0$. If b is not a critical value of f, by Theorem 1.9 with $\varepsilon < b$, there is an $\eta_1 \in C(E,E)$ such that $\eta_1(A_{b+\varepsilon}) \subset A_{b-\varepsilon}$. Choose $g \in \Gamma$ such that $\max_{g([0,1])} f \leq b + \varepsilon$. By 2° of Theorem 1.9, $\eta_1(0) = 0$, $\eta_1(e) = e$. Therefore $\eta_1 \circ g \in \Gamma$ and

$$b \leq \max_{u \in \eta_1 \circ g([0,1])} f(u) \leq b - \varepsilon$$

a contradiction. Hence the result.

Remark 3.10: (i) A function satisfying the hypotheses of Theorem 3.9 need neither be bounded from above or from below. E.g. for (2.15) with $n = 3$, $\lambda = 1$, and $g = u^3$,

$$f(u) = \frac{1}{2} \|u\|^2 - \frac{1}{4} \int_\Omega u^4 dx . \tag{3.11}$$

It is easy to see that $f \to -\infty$ along rays through the origin while one can construct a sequence (u_m) of spike-shaped functions which have $\max_\Omega |u_m(x)| \leq 1$ but $\|u_m\| \to \infty$. That (3.11) satisfies the hypotheses of Theorem 3.9 will be shown in Theorem 3.

(ii) Theorem 3.9 and most of the remainder of this section represents joint work of Ambrosetti and Rabinowitz [15].

There is a "dual" way to obtain a critical point of f satisfying (f_1), (f_2), and $(PS)^+$. Recall $\hat{A}_c = \{x \in E \mid f(x) \geq c\}$. Let Γ_* denote the set of homeomorphisms of E onto E with $h(0) = 0$ and $h(B_1) \subset \hat{A}_0$. Set $\Gamma_*^e = \{h \in \Gamma_* \mid h(S_1)$ separates 0 and $e\}$. Define

$$c = \sup_{h \in \Gamma_*^e} \inf_{v \in h(S_1)} f(v) .$$

<u>Theorem 3.12</u>: If f satisfies (f_1), (f_2), and $(PS)^+$, c is a critical value of f with $0 < \alpha \leq c \leq b < \infty$.

Proof: By (f_1), $h(u) = \rho u \in \Gamma_*^e$ so $\alpha \leq c$. Let $g \in \Gamma$ and $h \in \Gamma_*^e$. Since $h(S_1)$ separates 0 and e and $g([0,1])$ joins 0 and e, there exists a $w \in g([0,1]) \cap h(S_1)$. Therefore

$$\inf_{v \in h(S_1)} f(v) \leq f(w) \leq \max_{u \in g([0,1])} f(u) . \tag{3.13}$$

Since (3.13) is valid for all $h \in \Gamma_*^e$, $g \in \Gamma$, $c \leq b$. Then by Remark 1.23(ii) and 3°, 6° of Theorem 1.9, there is an $\varepsilon \in (0,\alpha)$ and a homeomorphism $\hat{\eta}_1$ of E onto E such that $\hat{\eta}_1(\hat{A}_{c-\varepsilon}) \subset \hat{A}_{c+\varepsilon}$. By 2° and 4° of that theorem, $\hat{\eta}_1(0) = 0$, $\hat{\eta}_1(e) = e$ and $\hat{\eta}_1 : \hat{A}_0 \to \hat{A}_0$. Choose $h \in \Gamma_*^e$ such that $\inf_{h(S_1)} f \geq c - \varepsilon$. Then $\hat{\eta}_1 \circ h : B_1 \to \hat{A}_0$. In addition, $\hat{\eta}_1 \circ h$ separates 0 and e. To verify this it suffices to show $e \notin \hat{\eta}_1 \circ h(B_1)$. But $\hat{\eta}_1(e) = e$ so $\hat{\eta}_1 \circ h(x) = e$ implies $h(x) = e$. Since $x \in B_1$ this is not possible. Thus $\hat{\eta}_1 \circ h \in \Gamma_*^e$. But by our choice of h,

$$\inf_{v \in \hat{\eta}_1 \circ h(S_1)} f(v) \geq c + \varepsilon$$

a contradiction.

Remark 3.14: (i) For $E \subset \mathbf{R}$, $b = c = \max_{x \in [0,e]} f(x)$. We do not know if they are equal in general.

(ii) Suppose we say a critical point u of f is a saddle point if every neighborhood of u contains points x, y at which $f(x) > f(u) > f(y)$. An open question is does K_b or K_c contain a saddle point. This will not be the case without further restrictions since e.g. if $E \subset \mathbf{R}^n$, b or c could easily correspond to a local maximum.

(iii) Theorems 3.9 and 3.12 are dual in the sense that for every $g \in \Gamma$ and for every $h \in \Gamma_*^e$, $g([0,1]) \cap h(S_1) \neq \phi$. In giving a variational characterization of critical values of f, there is a great deal of flexibility in choosing sets with respect to which to minimax. For example, if Γ is replaced by K, the set of compact connected sets meeting both 0 and e, and d is the inf max of f over elements of K, the proof of Theorem 3.9 shows d is a critical value of f. As a further illustration of this flexibility we give an improved version (modulo (PS)) of Theorem 3.12 replacing (f_1) by

(f_1') $f(0) = 0$ and 0 is a local minimum for f.

Let $W = \{\mathcal{O} \subset E \mid \mathcal{O}$ is open, $0 \in \mathcal{O}$, $e \notin \overline{\mathcal{O}}\}$. Thus if $\mathcal{O} \in W$, $\partial\mathcal{O}$ separates 0 and e. Define

$$w = \sup_{\mathcal{O} \in W} \inf_{u \in \partial\mathcal{O}} f(u) .$$

Theorem 3.15: If f satisfies (f_1'), (f_2), and (P.S.), w is a critical value of f. If $w = 0$, 0 is not an isolated critical point of f.

Proof: (f_1') implies $w \geq 0$. If $w > 0$, the proof of Theorem 3.12 carries over unchanged. Thus suppose $w = 0$. We can assume $\|e\| = 2$. If 0 is an isolated zero of f', by (f_1'), there is an $r > 0$ such that $f > 0$ in $B_r - \{0\}$. Since $f(e) = 0$, $r \leq 2$. Choose $N = B_{r/8}$ in Theorem 1.9. By that theorem there is an $\varepsilon > 0$ such that $\hat{\eta}_1(\hat{A}_{-\varepsilon} - N) \subset \hat{A}_\varepsilon$. Let $\mathcal{O} = B_{r/4}$ Then $\inf_{\partial\mathcal{O}} f = 0$ since $w = 0$. Hence $S_{r/4} \subset \hat{A}_{-\varepsilon} - N$. Therefore $\hat{\eta}_1(S_{r/4}) \subset \hat{A}_\varepsilon$. Moreover $\hat{\eta}_1(S_{r/4}) = \partial(\hat{\eta}_1(\mathcal{O}))$ since $\hat{\eta}_1$ is a homeomorphism and $\hat{\eta}_1(\mathcal{O}) \in W$. The latter statement follows from (1.12) since $\|\hat{\eta}_1(x) - x\| \leq 1$ for all $x \in E$ and therefore $\hat{\eta}_1(\mathcal{O}) \subset B_{1+r/4} \subset B_{3/2}$ so $\partial\hat{\eta}_1(\mathcal{O})$ separates 0 and e. However $\inf_{u \in \hat{\eta}_1(S_{r/4})} f(u) \geq \varepsilon > w$, a contradiction.

Remark 3.16: A slightly more careful argument shows if $w = 0$, the component of K_0 containing 0 meets S_r.

Next Theorem 3.9 and 3.12 will be applied to partial differential equations. Consider

$$(3.17) \qquad \begin{aligned} Lu &= p(x,u) \,, & x &\in \Omega \\ u &= 0 \,, & x &\in \partial\Omega \end{aligned}$$

where p satisfies (p_1), (p_2), (g_2), and

(p_5) $p(x,z)z^{-1} \to \infty$ as $z \to \infty$ uniformly in $x \in \overline{\Omega}$.

(p_6) There are constants $M > 0$ and $\theta \in [0, \frac{1}{2})$ such that $P(x,z) \leq \theta z p(x,z)$ for $|z| \geq M$.

The functional corresponding to (3.17) is

$$(3.18) \qquad f(u) = \frac{1}{2}\|u\|^2 - \int_\Omega P(x,u)\,dx \equiv \frac{1}{2}\|u\|^2 - \hat{f}(u)$$

defined and continuously differentiable on $E = W_0^{1,2}(\Omega)$ as earlier.

Theorem 3.19: If p satisfies $(p_1) - (p_2)$, $(p_5) - (p_6)$, and (g_2), then (3.17) possesses at least one nontrivial solution.

Proof: The hypotheses of Theorem 3.9 will be verified which then establishes the result. For (f_1), it suffices to show $\hat{f}(u) = o(\|u\|^2)$ at $u = 0$. This will be carried out for $n > 2$. (The $n = 2$ case is treated similarly.) Let $\varepsilon > 0$. By (p_2) there is a $\delta = \delta(\varepsilon) > 0$ such that $|P(x,z)| \leq \varepsilon|z|^2$ for $|z| \leq \delta$. Moreover by (g_2) there is an $\alpha_4 > 0$ such that $|P(x,z)| \leq \alpha_4|z|^{s+1}$ if $|z| \geq \delta$. Therefore by the Poincare inequality and (2.18),

$$(3.20) \qquad |\hat{f}(u)| \leq \int_\Omega (\varepsilon u^2 + \alpha_4|u|^{s+1})dx \leq \alpha_5(\varepsilon\|u\|^2 + \|u\|^{s+1}) .$$

Since by (g_2) we can assume $s > 1$, (3.20) implies (f_1).

Next to check (f_2), choose $u \in E$ with $\|u\| = 1$, $u > 0$ in Ω, and $\int_\Omega u^2 dx = \alpha_6$. Then

$$(3.21) \qquad f(Ru) = \frac{1}{2}R^2 - \hat{f}(Ru) .$$

By (p_5), for any $K > 0$, there exists a $\beta = \beta(K)$ such that $p(x,z) \geq Kz$ for $z \geq \beta$. Choosing $K = 4(\alpha_6)^{-1}$ yields

$$(3.22) \qquad f(Ru) \leq \frac{1}{2}R^2 + \alpha_7 - \int_{\{x\in\Omega \mid Ru(x)\geq\beta\}} \left(\int_\beta^{Ru(x)} p(x,t)\,dt\right)dx$$

$$\leq \frac{1}{2}R^2 + \alpha_8 - \int_{\{x\in\Omega \mid Ru(x)\geq\beta\}} \frac{K}{2}R^2u^2 dx .$$

For R sufficiently large,

$$\int_{\{x\in\Omega \mid Ru(x)\geq\beta\}} u^2 dx \geq \frac{1}{2}\alpha_6 .$$

Therefore

$$f(Ru) \leq \alpha_8 - \frac{1}{2}R^2 \tag{3.23}$$

so (f_2) is satisfied.

Lastly to verify $(PS)^+$, suppose $(u_m) \subset E$, $f(u_m) \leq d$ and $f'(u_m) \to 0$. Then

$$d \geq \frac{1}{2}\|u_m\|^2 - \alpha_9 - \int_{\{x\in\Omega \mid |u_m(x)|\geq M\}} P(x, u_m)dx \tag{3.24}$$

$$\geq \frac{1}{2}\|u_m\|^2 - \theta \int_\Omega u_m p(x, u_m)dx - \alpha_{10} .$$

Since $f'(u_m) \to 0$, for all $\varepsilon > 0$ there is a $k(\varepsilon)$ such that

$$|\langle f'(u_m), \varphi\rangle| = |(u_m, \varphi) - \int_\Omega p(x, u_m)\varphi\, dx| \leq \varepsilon\|\varphi\| \tag{3.25}$$

for all $m \geq k$ and all $\varphi \in E$. Choosing $\varepsilon = 1$ and $\varphi = u_m$ in (3.25) and combining with (3.24) yields

$$d \geq (\frac{1}{2} - \theta)\|u_m\|^2 - \theta\|u_m\| - \alpha_{10} . \tag{3.26}$$

Hence (u_m) is bounded in E. It now follows as in the proof of Theorem 3.4 that (u_m) possesses a convergent subsequence and the theorem is proved.

<u>Remark 3.27</u>: It is natural to question whether condition (g_2) used above and in Theorem 2.24 is necessary. The answer is in general, yes. This can be seen from an identity due to Pohozaev [27]. He showed if u is a

solution of

$$(3.28) \qquad -\Delta u = p(u) \;,\; x \in \Omega \qquad u = 0 \;,\; x \in \partial\Omega$$

then u satisfies

$$(3.29) \qquad 2n \int_\Omega P(u)dx + (2-n)\int_\Omega u\,p(u)dx = \int_{\partial\Omega} x\cdot\nu(x)|\nabla u|^2 ds$$

where $\nu(x)$ is the outward pointing normal to $\partial\Omega$. In particular if Ω is starshaped, i.e. $x\cdot\nu(x) \geq 0$ for $x \in \partial\Omega$,

$$(3.30) \qquad 2n \int_\Omega P(u)dx \geq (n-2)\int_\Omega u\,p(u)dx \,.$$

Taking the simple case $p(z) = z^s$ for $z \geq 0$ and p odd in (3.30) shows if $u \not\equiv 0$,

$$(3.31) \qquad n-2 \leq \frac{2n}{s+1} \quad \text{or} \quad s \leq \frac{n+2}{n-2} \,.$$

Possible equality in (3.31) can be excluded using (3.29) and the strong maximum principle. Thus (g_2) is necessary in general. However one can give examples where it is violated and nevertheless there are solutions. See [28].

Next we return to the case where f is even. For what follows it is assumed that E is an infinite dimensional Banach space. Analogous results are valid in finite dimensions with obvious modifications in their statement.

Suppose again $f \in C^1(E,\mathbb{R})$ and satisfies (f_1), $(PS)^+$, and

(f_3) f is even .

We replace (f_2) by a requirement that f be negative at infinity in an appropriate fashion:

(f_4) If X is a finite dimensional subspace of E, $X \cap \hat{A}_0$ is bounded.

Under the above hypotheses, which are essentially only conditions at 0 and infinity, we will show that f possesses infinitely many distinct pairs of critical points. As was mentioned earlier the main difficulty here is to find appropriate classes of sets with respect to which to minimax f. The sets $\overline{\gamma}_k$ used e.g. in Theorem 3.1 no longer suffice. For example the f in (3.11) satisfies the above hypotheses. However it is easy to see

$$\inf_{A \in \overline{\gamma}_k} \sup_{u \in A} f(u) = -\infty$$

while

$$\sup_{A \in \overline{\gamma}_k} \inf_{u \in A} f(u) = \infty$$

since f is neither bounded from above or from below. Even when one finds a class of sets for which a minimax argument succeeds, there is still the problem of obtaining an appropriate multiplicity result. We will show how to overcome these obstacles here.

Set $\Gamma^* = \{h \in \Gamma_* \mid h \text{ is odd}\}$ and

$$\Gamma_m = \{K \subset E \mid K \text{ is compact, symmetric (with respect to } 0), \text{ and for all } h \in \Gamma^*, \ \gamma(K \cap h(S)) \geq m\}$$

where $S = S_1$. Observe that if $h \in \Gamma^*$, $h(S) \subset E - \{0\}$ and is closed and symmetric. Therefore $K \cap h(S) \in \Sigma(E)$. The main properties of the sets Γ_m are contained in the following lemma.

Lemma 3.32: Let f satisfy (f_1) and (f_4). Then

1° $\Gamma_m \neq \phi$.

2° $\Gamma_{m+1} \subset \Gamma_m$.

3° If $K \in \Gamma_m$ and $Y \in \Sigma(E)$ with $\gamma(Y) \leq r < m$, then $\overline{K - Y} \in \Gamma_{m-r}$.

4° If φ is an odd homeomorphism of E onto E with $\varphi^{-1}(\hat{A}_0) \subset \hat{A}_0$, then $\varphi : \Gamma_m \to \Gamma_m$.

Proof:

1° Let X be an m dimensional subspace of E. If $K_R = X \cap B_R$, K_R is compact and symmetric. By (f_4) if R is sufficiently large, $K_R \supset X \cap \hat{A}_0$. If $h \in \Gamma^*$, $h(B_1) \subset \hat{A}_0$, so $K_R \supset X \cap h(B_1)$ and $K_R \cap h(S) = X \cap h(S)$. Moreover $h(0) = 0$ and $h(B_1)$ is a neighborhood of 0 in E, h being a homeomorphism. Therefore $X \cap h(B_1)$ is a symmetric bounded neighborhood of 0 in X with boundary contained in $X \cap h(S)$. Since X is isomorphic to $\mathbb{R}^m$, it follows from 2°, 3° of Lemma 1.1 and Theorem 1.2 that

$$(3.33) \qquad \gamma(K_R \cap h(S)) = \gamma(X \cap h(S)) \geq \gamma(\partial(X \cap h(B_1)) = m .$$

Since the right hand side of (3.33) can have at most genus m, actually we have equality in (3.33).

2° is trivial.

3° $\overline{K - Y}$ is compact and symmetric. If $h \in \Gamma^*$, then $\overline{K - Y} \cap h(S) = \overline{(K \cap h(S)) - Y}$ so by 5° of Lemma 1.1,

$$(3.34) \qquad \gamma(\overline{K - Y}) \cap h(S)) \geq \gamma(K \cap h(S)) - \gamma(Y) \geq m - r.$$

4° Let $K \in \Gamma_m$. Then $\varphi(K)$ is compact and symmetric and by 3° of Lemma 1.1,

$$\text{(3.35)} \quad \gamma(\varphi(K) \cap h(S)) = \gamma(K \cap \varphi^{-1} \circ h(S)) \quad \text{for } h \in \Gamma^* .$$

Since $h(S) \subset \hat{A}_0$, the hypotheses on φ imply $\varphi^{-1} \circ h \in \Gamma^*$. Hence the right hand side of (3.35) is not less than m and $\varphi : \Gamma_m \to \Gamma_m$.

Remark 3.36: The only role played by (f_4) above is in showing $\Gamma_m \neq \phi$. Hence it may be replaced by other hypotheses which serve the same purpose.

Theorem 3.37: Let $f \in C^1(E, \mathbb{R})$ and satisfy (f_1), (f_3) - (f_4), and $(PS)^+$. For $m \in \mathbb{N}$ define

$$b_m = \inf_{K \in \Gamma_m} \max_{u \in K} f(u) .$$

Then:

1° b_m is a critical value of f with $0 < \alpha \le b_m \le b_{m+1}$.

2° If $b_m = \dots = b_{m+r-1} \equiv b$, $\gamma(K_b) \ge r$.

3° $b_m \to \infty$ as $m \to \infty$.

Proof: Since $h(u) = \rho u \in \Gamma^*$, $K \cap h(S) = K \cap S_\rho \neq \phi$ for all $K \in \Gamma_m$. Therefore by (f_1), $\max_K f \ge \alpha$ for all $K \in \Gamma_m$ so $b_m \ge \alpha$. Since $\Gamma_{m+1} \subset \Gamma_m$, $b_{m+1} \ge b_m$. To show that b_m is a critical value of f_1 it suffices to prove the stronger multiplicity assertion 2°.

Suppose $\gamma(K_b) < r$. By 6° of Lemma 1.1, there is a $\delta > 0$ such that $\gamma(N_\delta(K_b)) < r$. Let $N = \text{int}\, N_\delta(K_b)$. Applying 3°, 6°, 7° of Theorem 1.9 shows there is an $\varepsilon \in (0, \alpha)$ and an odd homeomorphism

η_1 of E onto E such that $\eta_1(A_{b+\varepsilon} - N) \subset A_{b-\varepsilon}$. Choose $K \in \Gamma_{m+r-1}$ such that $\max_K f(u) \leq b + \varepsilon$. By 3° of Lemma 3.32, $K - N = \overline{K - N_\delta(K_b)} \equiv Q \in \Gamma_m$. Moreover $\eta_1^{-1}(\hat{A}_0) \subset \hat{A}_0$ by 4° of Theorem 1.9. Hence $\eta_1(Q) \in \Gamma_m$ by 4° of Lemma 3.32. But $b \leq \max_{\eta_1(Q)} f(u) \leq b - \varepsilon$, a contradiction.

Lastly to verify 3°, observe that by 1°, either (i) there is an $n \in \mathbb{N}$ such that $b_m = b_n$ for all $m \geq n$, (ii) (b_m) has a finite limit point, or (iii) $b_m \to \infty$ as $m \to \infty$. If (i) occurred, by 2°, $\gamma(K_{b_n}) = \infty$. However $(PS)^+$ implies K_{b_n} is compact and therefore by 6° of Lemma 1.1, $\gamma(K_{b_m}) < \infty$. To exclude (ii), suppose $b_m \to \overline{b} < \infty$. Let $\mathcal{K} = \{u \in E \mid b_1 \leq f(u) \leq \overline{b}$ and $f'(u) = 0\}$. Again by $(PS)^+$, $\mathcal{K}$ is compact. Suppose $\gamma(\mathcal{K}) = j$. By 6° of Lemma 1.1, there exists a $\delta > 0$ such that $\gamma(N_\delta(\mathcal{K})) = j$. Let $N = \text{int}\, N_\delta(\mathcal{K})$. By Theorem 1.9, there is an $\varepsilon > 0$ and an odd homeomorphism η_1 of E onto E such that $\eta_1(A_{\overline{b}+\varepsilon} - N) \subset A_{\overline{b}-\varepsilon}$. Let m be the smallest integer such that $b_m > \overline{b} - \varepsilon$. We can assume ε is small enough so that $m \geq 1$. Choose $K \in \Gamma_{m+j}$ so that $\max_K f \leq \overline{b} + \varepsilon$. By 3° and 4° of Lemma 3.32, $Q = K - N = \overline{K - N_\delta(\mathcal{K})} \in \Gamma_m$ and $\eta_1(Q) \in \eta_m$. Since $Q \subset A_{\overline{b}+\varepsilon} - N$, $\max_{\eta_1(Q)} f \leq \overline{b} - \varepsilon < b_m$, a contradiction.

As was the case with Theorem 3.9, a dual version of Theorem 3.37 obtains. This differs from the corresponding result in [15] in the choice of dual sets. Set

$$\Gamma_m^* = \{A \subset E \mid A \text{ is closed, symmetric, and } A \cap K \neq \phi \text{ for all } K \in \Gamma_m\}.$$

<u>Lemma 3.38</u>:

1° $\Gamma_m^* \neq \phi$.

2° $\Gamma_{m+1}^* \supset \Gamma_m^*$.

3° If $A \in \Gamma_m^*$ and $Y \in \Sigma(E)$ with $\gamma(Y) \leq r$, then $\overline{A - Y} \in \Gamma_{m+r}^*$.

4° If ψ is an odd homeomorphism of E onto E with $\psi(\hat{A}_0) \subset \hat{A}_0$, then $\psi : \Gamma_m^* \to \Gamma_m^*$.

Proof:

1° Since $\gamma(h(S) \cap K) \geq m$ for all $h \in \Gamma^*$ and $K \in \Gamma_m$, $h(S) \in \Gamma_m^*$.

2° If $A \in \Gamma_m^*$, $A \cap K \neq \phi$ for all $K \in \Gamma_m$. Since $\Gamma_{m+1} \subset \Gamma_m$, $A \cap K \neq \phi$ for all $K \in \Gamma_{m+1}$. Therefore $A \in \Gamma_{m+1}^*$.

3° We must show $\overline{A - Y} \cap K \neq \phi$ for all $K \in \Gamma_{m+r}^*$. As in Lemma 3.32, $\overline{A - Y} \cap K = \overline{(A \cap K) - Y} = \overline{(K \cap A) - Y} = \overline{K - Y} \cap A$. Since $A \in \Gamma_m^*$ and $\overline{K - Y} \in \Gamma_m$ by 3° of Lemma 3.32, this last set is nonempty.

4° Let $A \in \Gamma_m^*$ and $K \in \Gamma_m$. Then $\psi(A) \cap K \neq \phi$ is equivalent to $A \cap \psi^{-1}(K) \neq \phi$. By 4° of Lemma 3.32, $\psi^{-1} : \Gamma_m \to \Gamma_m$. Hence the result.

<u>Theorem 3.39</u>: Under the hypotheses of Theorem 3.37, if

$$c_m = \sup_{A \in \Gamma_m^*} \inf_{u \in A} f(u) \quad , \quad m \in \mathbb{N} ,$$

then 1° c_m is a critical value of f with $\alpha \leq c_m \leq b_m$ and $c_m \leq c_{m+1}$.

2° If $c_m = \ldots = c_{m+r-1} \equiv c$, $\gamma(K_c) \geq r$.

3° $c_m \to \infty$ as $m \to \infty$.

Proof: The proof is essentially the same as that of Theorem 3.37 with Lemma 3.32 replaced by Lemma 3.38 and will be omitted.

Remark 3.40: We do not know if $c_m = b_m$. Note also that one can define $\Gamma_m^{**} = \{K \subset E \mid K$ is closed, symmetric, and $K \cap A \neq \phi$ for all $A \in \Gamma_m^{*}\}$, $\Gamma_m^{***} = \ldots$, and define critical values of f by a corresponding minimax. Whether anything new is obtained by this process or whether the sets Γ_m, Γ_m^{**}, etc. eventually coincide, i.e. whether this duality stabilizes, is not known.

As an application of Theorems 3.37 and 3.39, we have

Theorem 3.41: Suppose $p(x, z)$ satisfies the hypotheses of Theorem 3.19 and is odd in z. Then (3.17) possesses infinitely many distinct pairs of solutions.

Proof: We need only verify the hypotheses of Theorem 3.37. (f_1) and $(PS)^+$ have already been verified in the proof of Theorem 3.19 and (f_3) is trivial. Lastly (f_4) follows on noting that the argument establishing (f_2) in Theorem 3.19 is uniform on finite dimensional subspaces of E.

Remark 3.42: Versions of Theorem 3.41 have been obtained under more restrictive hypotheses by Coffman (for an analogous integral equation) [17] and by Hempel [29]. Their assumptions enable them to put (3.17) into the setting of §2. They do this by observing that any solution of (3.17) satisfies

$$(3.43) \qquad \langle f'(u), u\rangle = \|u\|^2 - \int_\Omega p(x, u)u \, dx \equiv g(u) = 0 .$$

Under their conditions aside from $u = 0$, $g(u) = 0$ defines a manifold M radially homeomorphic to a sphere and $\langle g'(u), u\rangle \neq 0$ on this manifold. Thus the critical points of $f|_{g=0}$ satisfy

$$f'(u) - \lambda g'(u) = 0 \tag{3.44}$$

where λ is an appropriate Lagrange multiplier. Therefore

$$\langle f'(u), u\rangle = \lambda\langle g'(u), u\rangle \ . \tag{3.45}$$

Since $\langle g'(u), u\rangle \neq 0$ on M, (3.43) and (3.45) imply $\lambda = 0$. Hence (3.44) shows $f'(u) = 0$ so u is a critical point of f in E. Actually the problem of finding critical points of $f|_{g=0}$ is not quite of the form treated in §2 since M is not bounded. However the techniques of §2 work on this manifold anyway.

Results in the spirit of Theorems 3.9, 3.12, 3.37, and 3.39 can be obtained by relaxing (f_1) and (f_4). For work in this direction and applications to partial differential equations, see [15].

4. Variational methods for bifurcation problems.

Let E be a real Hilbert space and Ω a neighborhood of 0 in E. Suppose $L, H \in C(\Omega, E)$ with L linear and $H(u) = o(\|u\|)$ at $u = 0$. Consider the equation

$$Lu + H(u) = \lambda u \tag{4.1}$$

with a solution being a pair $(\lambda, u) \in \mathbf{R} \times E$. Thus (4.1) possesses the line of trivial solutions $\{(\lambda, 0) \mid \lambda \in \mathbf{R}\}$. We say $(\mu, 0)$ is a bifurcation point for (4.1) if every neighborhood of $(\mu, 0)$ contains nontrivial solutions.

Bifurcation phenomena arise in many physical contexts and have been extensively studied. A necessary condition for $(\mu, 0)$ to be a bifurcation point is that μ belong to the spectrum of L for otherwise $(L - \lambda I)$ is invertible for all λ near μ and

$$(4.2) \qquad u = -(L - \lambda I)^{-1} H(u) .$$

But since $H(u) = o(\|u\|)$ at $u = 0$, (4.2) can have no small nonzero solutions. This necessary condition is not sufficient as simple examples show. E. g. for $E = \mathbb{R}^2$, $u = (x, y)$, $L = I$, and $H = (-y^3, x^3)$, $\mu = 1$ is the only point in the spectrum of L but multiplying the first equation in (4.1) by y, the second by x and subtracting shows $(1, (0, 0))$ is not a bifurcation point.

Suppose however that $Lu + H(u) = f'(u)$ where $f \in C^2(\Omega, \mathbb{R})$. Then under additional mild conditions, the necessary condition is sufficient. The first result of this type is due to Krasnoselski [3] who treated the case $f \in C^1(\Omega, \mathbb{R})$, f' compact and uniformly differentiable near 0, and $f''(0)$ exists. His methods were variational and related to the arguments of §2. Several people have made improvements of this result. See e.g. [4-7], [11], [30], [31]. Relatively recently this problem has been treated independently by Böhme [6] and Marino [7] using an elementary variational argument. They proved

<u>Theorem 4.3:</u> If $f \in C^2(\Omega, \mathbb{R})$ and μ is an isolated eigenvalue of L of finite multiplicity, then $(\mu, 0)$ is a bifurcation point for (4.1). Moreover (4.1) possesses at least two distinct one parameter families of solutions $(\lambda(\varepsilon), u(\varepsilon))$ having $\|u(\varepsilon)\| = \varepsilon$ and $\lambda(\varepsilon) \to \mu$ as $\varepsilon \to 0$.

Actually Böhme and Marino allow a more general right hand side in (4.1). In addition Marino assumed L to be compact but only uses this in his argument to get μ to be an isolated eigenvalue of finite multiplicity. Recently McLeod and Turner [32] have weakened the requirement that f be C^2 to C^1 with the first derivative being Lipschitz continuous with a small Lipschitz constant. Berger [33] has permitted a more general λ dependence in (4.1).

We will prove Theorem 4.3 below by a slight modification of the argument used in [6] and [7]. One disadvantage of Theorem 4.3 is that it offers no information on the number of small solutions of (4.1) near $(\mu, 0)$ as a function of λ. We will also indicate some results in this direction which depend on the minimax arguments of § 3.

Proof of Theorem 4.3: The proof consists of three steps. (A) First the problem is reduced to a finite dimensional one by a standard reduction. (B) A second reduction is made eliminating λ. (C) The solutions are obtained by extremizing f on a finite dimensional compact manifold.

(A) A Lyapunov-Schmidt reduction.

Let X denote the null space of $(L - \mu I)$ having e.g. dimension n. Then $E = X \oplus X^{\perp}$ and $u \in E$ implies $u = v + w$, $v \in X$, $w \in X^{\perp}$. Let $P, P^{\perp}$ denote respectively the orthogonal projectors of E onto X and $X^{\perp}$. Then the equation

$$f'(u) \equiv Lu + H(u) = \lambda u \tag{4.4}$$

is equivalent to the pair of equations

$$\text{(4.5)} \qquad Pf'(u) = Lv + PH(v+w) = \lambda v$$

$$\text{(4.6)} \qquad P^{\perp}f'(u) = Lw + P^{\perp}H(v+w) = \lambda w .$$

Set $\lambda = \mu + \beta$ and define

$$\text{(4.7)} \qquad F(\beta, v, w) = (L - (\mu + \beta)I)w + P^{\perp}H(v+w) .$$

Then F is continuously differentiable near $(0,0,0)$ in $\mathbf{R} \times X \times X^{\perp}$, $F(0,0,0) = 0$, and $F_w(0,0,0) = L - \mu I$, an isomorphism from $X^{\perp}$ to $X^{\perp}$. Hence by the implicit function theorem, the zeroes of F near $(0,0,0)$ are given precisely by $w = \varphi(\beta, v)$ with $\varphi(0,0) = 0$ and φ continuously differentiable near $(0,0)$ in $\mathbf{R} \times X$. Thus to solve (4.4) near $(\mu, 0)$, it suffices to solve the finite dimensional problem (4.5) with $w = \varphi(\beta, v)$ or equivalently since $Lv = \mu v$,

$$\text{(4.8)} \qquad \beta v = PH(v + \varphi(\beta, v)) .$$

This finite dimenstional reduction is usually called the <u>method of Lyapunov-Schmidt</u>.

For what follows, some estimates are required. Since $(L - \lambda I)|_{X^{\perp}}$ is an isomorphism on $X^{\perp}$ for all λ near μ, (4.6) implies

$$\text{(4.9)} \qquad \|\varphi(\beta, v)\| \le \alpha_1 \|H(v + \varphi(\beta, v))\| \le o(\|v\| + \|\varphi(\beta, v)\|)$$

and therefore

$$\text{(4.10)} \qquad \|\varphi(\beta, v)\| = o(\|v\|) \quad \text{at} \quad v = 0$$

uniformly in small β. Differentiating (4.6) with respect to β yields

$$(L - \lambda I + P^{\perp}H'(v + \varphi(\beta, v)))\frac{\partial \varphi}{\partial \beta} = \varphi(\beta, v) . \tag{4.11}$$

Since $H'(0) = 0$ and v is near 0, it follows from (4.10) that

$$\left\|\frac{\partial \varphi}{\partial \beta}(\beta, v)\right\| = o(\|v\|) \text{ at } v = 0 \tag{4.12}$$

uniformly in small β.

(B) A second reduction.

Taking the inner product of (4.8) with v shows for $v \neq 0$,

$$\beta = (H(v + \varphi(\beta, v)), v)\|v\|^{-2} . \tag{4.13}$$

Set $G(\beta, v) = \beta - (H(v + \varphi(\beta, v)), v)\|v\|^{-2}$ for $v \neq 0$ and $G(\beta, 0) = \beta$. Then it readily follows with the aid of (4.10) and (4.12) that G is continuous near $(0,0)$ in $\mathbb{R} \times X$, it is continuously differentiable with respect to β, $G(0,0) = 0$, and $\frac{\partial G}{\partial \beta}(0,0) = 1$. It is also continuously differentiable with respect to v away from $v = 0$. Hence by a slightly stronger version of the implicit function theorem (or by the contracting mapping theorem) the zeroes of G near $(0,0)$ are given by $\beta = \psi(v)$ with $\psi(0) = 0$ and ψ continuous near 0 and differentiable in a deleted neighborhood of 0.

Let $\chi(v) = \varphi(\psi(v), v)$. Then χ is continuous near 0 in X and differentiable in a deleted neighborhood of 0. By (4.10), $\|\chi(v)\| = o(\|v\|)$ at $v = 0$. We also need an estimate for $\chi'(v)z$ near $v = 0$. Setting $w = \chi(v)$ and $\lambda = \mu + \psi(v)$ in (4.6), replacing v by $v + tz$ and differentiating with respect to t at $t = 0$ yields:

$$(L - (\mu + \psi)I + P^{\perp}H'(v + \chi))\chi'(v) = (\psi', z)\chi - P^{\perp}H'(v + \chi)z . \tag{4.14}$$

Similarly from (4.13) for $v \neq 0$,

$$(4.15) \qquad (\psi', z) = [(H'(v + \chi)(z + \chi'(v)z), v) + (H(v + \chi), z)$$

$$- 2(H(v + \chi), v)(v, z) \|v\|^{-2}] \|v\|^{-2} .$$

Combining (4.14) - (4.15) shows

$$(4.16) \qquad \|\chi'(v)z\| \leq \alpha_1[(\frac{\|H'(v + \chi)\|}{\|v\|} (\|z\| + \|\chi'(v)z\|)$$

$$+ \frac{3\|H(v + \chi)\| \, \|\chi\|}{\|v\|^2}) \|\chi\| + \|H'(v + \chi)\| \, \|z\|] .$$

Since $\|H'(v + \chi)\|$, $|\psi(v)| = o(1)$ and $\|H(v + \chi)\|$, $\|\chi(u)\| = o(\|v\|)$ at $v = 0$, it follows from (4.16) that

$$(4.17) \qquad \|\chi'(v)z\| = o(1) \|z\| \text{ at } v = 0 .$$

Hence χ is C^1 near 0.

(C) A finite dimensional variational problem.

Let U be a sufficiently small neighborhood of 0 in X for ψ and χ to be defined and let $M = \{v + \chi(v) \mid v \in U\}$. (4.17) shows M is a C^1 n manifold in E. Let $\varepsilon > 0$ and $N = N(\varepsilon) = M \cap S_\varepsilon$. Then for ε sufficiently small, N is a compact $C^1(n - 1)$ manifold in E.

Consider $f|_N$. This possesses at least two distinct critical points, namely the points at which the maximum and minimum of $f|_N$ are achieved. We will show for any critical point of $f|_N$, (4.4) is satisfied. Since $\psi(v) \to 0$ as $v \to 0$, the proof will be complete.

Let TZ_x denote the tangent space to a C^1 manifold Z at x. Since $N = M \cap S_\varepsilon$, $TN_x = TM_x \cap T(S_\varepsilon)_x$. Clearly $T(S_\varepsilon)_x = \{z \in E \mid (z, x) = 0\}$.

Therefore $TN_x = \{z \in TM_x \mid (z,x) = 0\}$.

Suppose u is a critical point of $f|_N$ i.e. $(f'(u),z) = 0$ for all $z \in TN_u$. Then

$$(4.18) \qquad (f'(u) - \varepsilon^{-2}(f'(u),u)u, z) = 0 \quad \text{for all} \quad z \in \text{span}\{u, TN_u\} \ .$$

We must relate (4.18) to (4.6), (4.13) which determine w and β. Since $u \in M$, $u = v + \chi(v)$ satisfies (4.6). Taking the inner product of (4.6) with $w = \chi(v)$ yields

$$(4.19) \qquad (f'(u),w) = (\mu + \psi(v)) \|w\|^2 \ .$$

Since u satisfies (4.13) and $\mu\|v\|^2 = (Lv,v)$, (4.13) implies

$$(4.20) \qquad (f'(u),v) = (\mu + \psi(v)) \|v\|^2 \ .$$

Combining (4.19) and (4.20) shows

$$(4.21) \qquad (f'(u),u) = (\mu + \psi(v)) \|u\|^2 = \lambda\varepsilon^2 \ .$$

Thus if u also satisfies (4.18), that equation can be rewritten as

$$(4.22) \qquad (f'(u) - \lambda u, z) = 0 \quad \text{for all} \quad z \in \text{span}\{u, TN_u\} \ .$$

From (4.6) we also have $f'(u) - \lambda u$ is orthogonal to $X^\perp$ and hence to $W \equiv \text{span}\{u, TN_u, X^\perp\}$.

To complete the proof we will show $W = E$ so (4.4) is satisfied. Since TM_u is the range of the Frechet derivative of the map $v \to u = v + \chi(v)$, $TM_u = \{x + \chi'(v)x \mid x \in X\}$. Therefore if $y \in E$,

$$(4.22) \qquad y = (P + \chi'(v)P)y + (P^\perp - \chi'(v)P)y \in \text{span}\{TM_u, X^\perp\} \ .$$

Suppose $v_1, \dots, v_{n-1}$ form a basis for TN_u and when supplemented by v_n form a basis for TM_u. Hence $(v_n, u) \neq 0$ or TN_u would be n dimensional. Since $v + \chi'(v)v \in TM_u$,

$$v + \chi'(v)v = \sum_1^n \beta_i v_i . \tag{4.23}$$

In (4.23), $\beta_n \neq 0$ for otherwise, this vector lies in TN_u so

$$(v + \chi'(v)v, u) = 0 = \|v\|^2 + (\chi'(v)v, \chi(v)) . \tag{4.24}$$

But from (4.12) and (4.17), the right hand side of (4.24) is $\|v\|^2 + o(\|v\|^2)$ which vanishes only for $v = 0$ if v is small. Thus setting $y = v$ in (4.22), v_n can be determined in terms of v and elements of TN_u and $X^\perp$, i.e. $v_n \in \text{span}\{v, TN_u, X^\perp\} = \text{span}\{u, TN_u, X^\perp\} = W$. Consequently (4.22) with an arbitrary y shows $y \in W$, i.e. $W = E$ and the proof is complete.

<u>Remark 4.25:</u> If f is even, φ and χ are odd in v and $f|_N$ is even. Hence the theory of §2 can be applied to show $f|_N$ possesses at least n distinct pairs of critical points.

As an application of Theorem 4.3, consider (3.2) with p satisfying (p_2) and e.g.

$$p(x,z) \in C^1(\overline{\Omega} \times [-M,M], \mathbb{R}) . \tag{p_1'}$$

Extend p to $\overline{\Omega} \times \mathbb{R}$ in a C^1 fashion so that p_z is globally bounded. Then it readily follows that

$$f(u) = \int_\Omega \left(\frac{1}{2} au^2 + P(x,u)\right) dx$$

is C^2 in a neighborhood of 0 in $E = W_0^{1,2}(\Omega)$. Then Theorem 4.3 obtains at all eigenvalues of (3.3) and we conclude that every eigenvalue of (3.3) corresponds to a bifurcation point for (3.2) with the modified p. However elliptic regularity theory implies $\max_{\Omega} |u(x)| \to 0$ as $\|u\| \to 0$ for these bifurcation solutions so in fact for $\|u\|$ small (λ, u) satisfies the original equation.

We conclude this section by remarking that there is another way to prove the first assertion of Theorem 4.3 more in the spirit of §3, in particular Theorem 3.9. This proof also gives more information on the number of solutions of (4.1) as a function of λ near $(\mu, 0)$. The proof will not be given here (see [34]) however we state the result. Let $\mathcal{S}$ denote the closure of the set of nontrivial solutions of (4.1) in $\mathbf{R} \times \overline{\Omega}$.

Theorem 4.26: If $f \in C^2(\Omega, \mathbf{R})$ and μ is an isolated eigenvalue of L of finite multiplicity, then $(\mu, 0)$ is a bifurcation point for (4.1). Moreover either (i) $(\mu, 0)$ is not an isolated solution of (4.1) in $\{\mu\} \times \Omega$ (ii) the projection of $\mathcal{S}$ on $\mathbf{R}$ contains a (nonempty) open interval which includes μ as an interior point, or (iii) the projection of $\mathcal{S}$ on $\mathbf{R}$ contains a (nonempty) one sided neighborhood of μ such that for all $\lambda \in \Lambda$, (4.1) possesses at least two distinct solutions.

Remark 4.27: (i) Note that (i) above occurs if $H \equiv 0$.

(ii) A comparison of Theorem 4.3 and 4.26 suggests there is a more optimal result which contains both of them. The negation of (i) in Theorem 4.26 is essentially a transversality assumption and transversality is intrinsic in the proof of Theorem 4.3 so perhaps this is the key to a better result.

(iii) Stronger assertions can be made if f is even. See e.g. [35] or [34].

5. The interplay of variational and continuation methods.

In this section we shall show for a particular class of partial differential equations how variational methods may be used in conjunction with continuation arguments to obtain further information about the number of solutions the equation possesses. These results represent joint work with Crandall [36].

To begin, consider

$$\text{(5.1)} \qquad \begin{aligned} Lu &= \lambda h(x,u) \ , \quad && x \in \Omega \\ u &= 0 \ , && x \in \partial\Omega \end{aligned}$$

where h satisfies

(h_1) $h \in C^3(\overline{\Omega} \times \mathbf{R}, \mathbf{R}^+)$

(h_2) $h(x,0) > 0$, $h_z(x,0) > 0$, and $h_{zz}(x,z) > 0$ for $x \in \Omega$, $z > 0$.

As usual a solution of (5.1) means a pair $(\lambda,\mu) \in \mathbf{R} \times C^2(\Omega)$. Let $P^+ = \{u \in C^2(\Omega) \,|\, u > 0 \text{ in } \Omega, \frac{\partial u}{\partial \nu} < 0 \text{ on } \partial\Omega\}$. (h_2) implies any solution (λ, u) of (5.1) with $\lambda > 0$ has $u \in P^+$. Indeed if $A = \{x \in \Omega \,|\, u(x) < 0\}$, then $Lu > 0$ in A and $u = 0$ on ∂A. The maximum principle implies $u \equiv 0$ in A so $A = \phi$. The strong maximum principle then shows $u \in P^+$. Under weaker assumptions than (h_1) - (h_2), it follows from arguments involving topological degree that (5.1) possesses a component of solutions meeting (0,0) and unbounded in $\mathbf{R}^+ \times P^+$ [19]. However we are interested in more information on the number of solutions as a function of λ.

Unlike our earlier applications, (5.1) possesses no line of trivial solutions. However $(0,0)$ is a solution of (5.1) and a simple continuation argument gives a family of solutions in $\mathbb{R}^+ \times \mathbb{P}^+$. Let $\alpha \in (0,1)$ and let $C^{k,\alpha}(\Omega)$ denote the space of functions which are k times continuously differentiable in $\bar{\Omega}$ and whose k^{th} derivatives are Hölder continuous with exponent α. We take the usual norm in $C^{k,\alpha}(\Omega)$, $C^{0,\alpha}(\Omega) \equiv C^{\alpha}(\Omega)$ and $C_0^{2,\alpha}(\Omega) = \{u \in C^{2,\alpha}(\Omega) \mid u = 0 \text{ on } \partial\Omega\}$. (h_1) implies solutions lie in $\mathbb{R} \times C_0^{2,\alpha}(\Omega)$. Linearizing (5.1) about $u = 0$ gives a linear eigenvalue problem.

$$
(5.2) \qquad \begin{aligned} Lv &= \lambda h_z(x,0)v \ , && x \in \Omega \\ v &= 0 \ , && x \in \partial\Omega \ . \end{aligned}
$$

Let λ_1 denote the smallest eigenvalue of (5.2). As is well known, it is positive, simple, and possesses an eigenfunction $v_1 \in P^+$.

Lemma 5.3: There exists $\bar{\lambda} \in (0,\lambda_1)$ which is maximal with respect to the existence of a curve $\{(\lambda, u(\lambda)) \mid \lambda \in [0,\bar{\lambda})\}$ of solutions of (5.1) satisfying:

(i) The mapping $\lambda \to u(\lambda)$, $[0,\bar{\lambda}) \to C_0^{2,\alpha}(\Omega)$ is continuous.

(ii) The mapping $v \to Lv - \lambda h_z(\cdot, u(\lambda)(\cdot))v$, $C_0^{2,\alpha}(\Omega) \to C^{\alpha}(\Omega)$ is invertible for $\lambda \in [0,\bar{\lambda})$.

Proof: Let $\Psi(\lambda, u) = Lu - \lambda h(x,u)$. Then it is easily seen that $\Psi \in C^1(\mathbb{R} \times C_0^{2,\alpha}(\Omega), C^{\alpha}(\Omega))$, $\Psi(0,0) = 0$, and $\Psi_u(0,0) = L$ which is an isomorphism from $C_0^{2,\alpha}(\Omega)$ to $C^{\alpha}(\Omega)$. By the implicit function theorem the solutions $(\lambda, u(\lambda))$ of (5.1) near $(0,0)$ are given by a C^1 curve passing through $(0,0)$. This curve can be continued using the implicit function theorem to a maximal positive interval $[0,\bar{\lambda})$ in which (ii) is satisfied. To see that $\bar{\lambda} < \lambda_1$, observe that $h_z(x,0) \leq h_z(x,u(\lambda))$ via

(h_2) and $u(\lambda) \in P^+$. Hence the smallest eigenvalue of (5.2) is larger than the smallest eigenvalue $\mu_1(\lambda)$ of

$$(5.4) \qquad \begin{aligned} Lw &= \mu h_z(x,u(\lambda))w \;, & x \in \Omega \\ w &= 0 \;, & x \in \partial\Omega \;. \end{aligned}$$

Hence $\lambda < \mu_1(\lambda) \leq \lambda_1$.

Remark 5.5: It can also be shown that $u(\lambda)(x)$ is nondecreasing as a function of λ and for fixed $\lambda \in (0,\overline{\lambda})$, $u(\lambda)$ is pointwise the smallest solution of (5.1) in P^+.

Next we will show under additional hypotheses on h at infinity there are at least two solutions of (5.1) for each $\lambda \in (0,\overline{\lambda})$.

Theorem 5.6: Suppose h satisfies $(h_1) - (h_2)$, (g_2), and $(p_5) - (p_6)$. Then for all $\lambda \in (0,\overline{\lambda})$, (5.1) possesses at least two distinct solutions $(\lambda,u(\lambda))$, $(\lambda,\overline{u}(\lambda))$ with $u(\lambda),\overline{u}(\lambda) \in P^+$.

Proof: Since the values of h for $z < 0$ are irrelevant here, redefine h in $\overline{\Omega} \times (-\infty,0)$ so that $h > 0$ in $\overline{\Omega} \times \mathbb{R}$, satisfies (g_2) and (p_4), and is e.g. locally Lipschitz continuous. Fix $\lambda \in (0,\overline{\lambda})$ and define

$$H(x,z) = \int_0^z h(x,t)\,dt\,,\; W(x,z) = H(x,z+u(\lambda)(x)) - H(x,u(\lambda)(x)) - h(x,u(\lambda)x)z$$

and

$$(5.7) \qquad f(u) = \frac{1}{2}\,\|u\|^2 - \lambda \int_\Omega W(x,u)\,dx\;.$$

Again $E = W_0^{1,2}(\Omega)$ and $f \in C^1(E,\mathbb{R})$. Then $f(0) = 0$ and 0 is a critical point of f corresponding to the solution $(\lambda,u(\lambda))$ of (5.1). We claim that f satisfies the hypotheses of Theorem 3.9. Assuming this for the moment, it follows that f possesses a second critical point $\overline{u}(\lambda)$ and $(\lambda, \overline{u}(\lambda))$ will be a second positive solution of (5.1). (That $\overline{u} \in P^+$ is a consequence of earlier remarks.)

It remains to verify the hypotheses of Theorem 3.9. $(PS)^+$ and (f_2) follow from (g_2), (p_5) - (p_6) as was shown in the proof of Theorem 3.19. To check (f_1), for convenience let $n > 2$. By (g_2),

$$W(x,z) = \frac{1}{2} h_z(x, u(\lambda)(x)) z^2 + R(x,z) \tag{5.8}$$

where the remainder term R is such that for all $\varepsilon > 0$, there is a $\delta(\varepsilon) > 0$ with

$$\begin{aligned} |R(x,z)| &\leq \varepsilon z^2 \quad , \quad |z| \leq \delta(\varepsilon) \\ &\leq \alpha_\varepsilon |z|^{s+1} \quad , \quad |z| \geq \delta(\varepsilon) \end{aligned} \tag{5.9}$$

where $s > 1$. Substituting in (5.7) yields

$$f(u) \geq \frac{1}{2}(\|u\|^2 - \lambda \int_\Omega h_z(x, u(\lambda)) u^2 dx) - \varepsilon \int_\Omega u^2 dx - \alpha_\varepsilon \int_\Omega |u|^{s+1} dx. \tag{5.10}$$

Since $\Psi_u(\lambda, u(\lambda))$ has positive spectrum for $\lambda = 0$ and is an isomorphism for all $\lambda \in (0, \overline{\lambda})$, it follows that the smallest eigenvalue of $\Psi_u(\lambda, u(\lambda))$ is positive. Therefore there is a constant $\alpha_1 > 0$ (and independent of ε) such that

$$\|w\|^2 - \lambda \int_\Omega h_z(x, u(\lambda)) w^2 dx \geq \alpha_1 \|w\|^2 \tag{5.11}$$

for all $w \in E$. Moreover by the Poincare and Sobolev inequalities

$$\|w\|^2 \geq \alpha_2 \int_\Omega w^2 dx \; ; \quad \alpha_3 \|w\| \geq (\int_\Omega |w|^{s+1} dx)^{\frac{1}{s+1}} \tag{5.12}$$

for $w \in E$. Substituting (5.11) - (5.12) with $w = u$ in (5.10) shows

$$f(u) \geq (\frac{1}{2}\alpha_1 - \frac{\varepsilon}{\alpha_2})\|u\|^2 - \alpha_3^{s+1}\|u\|^{s+1}. \tag{5.13}$$

Choosing ε so that $\varepsilon < \frac{1}{2}\alpha_1\alpha_2$ then gives (f_1) and the proof is complete.

References

[1] Ljusternik, L. A., Topologische Grundlagen der allgemainen Eigenwerttheorie, Monatsch. Math. Phys. 37, (1930), 125-130.

[2] Palais, R. S., Critical point theory and the minimax principle, Proc. Sym. Pure Math., 15, A.M.S., Providence, R.I. (1970), 185-212.

[3] Karasnoselski, M. A., Topological Methods in the Theory of Nonlinear Integral Equations, Macmillan, New York, 1964.

[4] Marino, A. and G. Prodi, La teoria di Morse per gli spazi di Hilbert, Rend. Sem. Mat. Univ. Padova, 41, (1968), 43-68.

[5] Naumann, J., Variationsmethoden für Existenz und Bifurkation von Lösungen nichtlinearer Eigenwertprobleme I & II, Math. Nacht. 54, (1972), 285-296, 55 (1973), 325-344.

[6] Böhme, R., Die Lösung der Verzweigungsgleichungen für nichtlineare Eigenwertprobleme,Math. Z., 127, (1972), 105-126.

[7] Marino, A., La biforcazione nel caso variazionalle, Proc. Conference del semanario di Mathematica dell' Universetà di Bari, Nov. 1972, to appear.

[8] Ljusternik, L. A. and L. G. Schnirelmann, Topological Methods in the Calculus of Variations, Hermann, Paris, 1934.

[9] Vainberg, M. M., Variational Methods for the Study of Nonlinear Operators, Holden-Day, San Francisco, 1964.

[10] Schwartz, J. T., Nonlinear Functional Analysis, lecture notes, Courant Inst. of Math. Sc., New York, Univ. 1965.

[11] Fučik, S., J. Nečas, J. Souček, and V. Souček, Spectral Analysis of Nonlinear Operators, Lecture Notes in Mathematics #343, Springer Verlag, 1973.

[12] Palais, R., Ljusternik-Schnirelman theory on Banach manifolds, Topology 5, (1966), 115-132.

[13] Browder, F. E., Existence theorems for nonlinear partial differential equations, Proc. Sym. Pure Math. 16, A.M.S. Providence, (1970), 1-60.

[14] Browder, F. E., Nonlinear eigenvalue problems and group invariance, appearing in Functional Analysis and Related Fields, F. E. Browder, editor, Springer (1970), 1-58.

[15] Ambrosetti, A. and P. H. Rabinowitz, Dual variational methods in critical point theory and applications, J. Funct. Anal. 14, (1973), 349-381.

[16] Berger, M., Applications of global analysis to specific nonlinear eigenvalue problems. Rocky Mtn. Math. J. 3 (1973), 319-354.

[17] Coffman, C. V., A minimum-minimax principle for a class of nonlinear integral equations, J. Analyse Math. 22 (1969), 391-419.

[18] Connor, E. and E. E. Floyd, Fixed point free involutions and equivariant maps, Bul. A.M.S., 66 (1960), 416-441.

[19] Rabinowitz, P. H., Some aspects of nonlinear eigenvalue problems, Rocky Mtn. J. of Math., 3 (1973), 161-202.

[20] Clark, D. C., A variant of the Ljusternik-Schnirelmann Theory, Ind. Univ. Math. J., 22 (1972), 65-74.

[21] Amann, H., Ljusternik-Schnirelman Theory and nonlinear eigenvalue problems, Math. Ann., 199 (1972), 55-72.

[22] Fučik, S., J. Nečas, J. Souček, and V. Souček, Upper bounds for the number of critical levels for nonlinear operators in Banach spaces of the type of second order nonlinear partial differential equations, J. Func. Anal., 11 (1972), 314-344.

[23] Schwartz, J. T., Generalizing the Ljusternik-Schnirelman theory of critical points, Comm. Pure Appl. Math., 17 (1964), 307-315.

[24] Freedman, A., Partial Differential Equations, Holt, Rinehart, and Winston Inc., New York, 1969.

[25] Hempel, J. A., Multiple solutions for a class of nonlinear elliptic boundary value problems, Indiana Univ. Math. J., 20 (1971), 983-996.

[26] Ambrosetti, A., On the existence of multiple solutions for a class of nonlinear boundary value problems, Rend. Sem. Mat. Univ. Padova, 49 (1973), 195-204.

[27] Pohozaev, S. I., Eigenfunctions of the equation $\Delta u + \lambda f(u) = 0$, Sov. Math., 5 (1965), 1408-1411.

[28] Rabinowitz, P. H., Variational methods for nonlinear elliptic eigenvalue problems, Indiana Univ. Math. J., 23 (1974), 729-754.

[29] Hempel, J., Superlinear variational boundary value problems and nonuniqueness, thesis, Univ. of New England, Australia, 1970.

[30] Reeken, M., Stability of critical points under small perturbations, Part II: Analytic theory, Manus. Math., 8 (1973), 69-92.

[31] Reeken, M., Stability of critical values and isolated critical continua, Math. Report #79, Battelle Advanced Studies Center, Geneva, Switzerland, Nov. 1973.

[32] McLeod, B. and R. E. L. Turner, Bifurcation for nondifferentiable operators, to appear.

[33] Berger, M. S., Bifurcation theory and the type numbers of Marston Morse, Proc. Nat. Acad. Sc., 69 (1972), 1737-1738.

[34] Rabinowitz, P. H., A bifurcation theorem for potential operators, to appear.

[35] Clark, D. C., Eigenvalue bifurcation for odd gradient operators, to appear.

[36] Crandall, M. G. and P. H. Rabinowitz, Some continuation and variational methods for positive solutions of nonlinear elliptic eigenvalue problems, to appear Arch. Rat. Mech. Anal.

Department of Mathematics
University of Wisconsin
Madison, Wisconsin 53706

CENTRO INTERNAZIONALE MATEMATICO ESTIVO

(C.I.M.E.)

EXISTENCE OF SOLUTIONS TO THE HARTREE-FOCK EQUATIONS

M. REEKEN

Corso tenuto a Varenna dal 16 al 25 giugno 1974

EXISTENCE OF SOLUTIONS TO THE HARTREE-FOCK EQUATIONS

by

M. Reeken

In Quantum Chemistry the following system of equations is considered:

$$(1)\quad \begin{aligned} &\left(-\tfrac{1}{2}\Delta - \frac{Z}{|x|}\right)\bar{u}(x) + \bar{u}(x)\int\frac{\bar{u}^2(y)}{|x-y|}dy - \int\frac{\left(\bar{u}(x)\cdot\bar{u}(y)\right)\bar{u}(y)}{|x-y|}dy = \bar{\Lambda}\cdot\bar{u}(x) \\ &\int u_i(x)u_j(x)dx = \delta_{ij} \end{aligned}$$

$\bar{u}(x) = \left(u_1(x),\ldots,u_Y(x)\right)$ a vector of square integrable functions

$\bar{\Lambda}$ a real symmetric matrix

$$\bar{u}(x)\cdot\bar{u}(y) = \sum_1^Y u_i(x)u_i(y) \quad .$$

These equations go by the name of Hartree-Fock equations. There is also a single equation called Hartree equation:

$$(2)\quad \begin{aligned} &\left(-\tfrac{1}{2}\Delta - \frac{Z}{|x|}\right)u(x) + u(x)\int\frac{u^2(y)}{|x-y|}dy = \lambda u(x) \\ &\int u^2(x)dx = 1 \quad . \end{aligned}$$

Both are the Euler-Lagrange equations for the critical points of the quadratic form $(\psi,H\psi)$ under different constraints. Here we denote as usual the Schrödinger operator of an atom of nuclear charge Z and electron number Y by H .

$$H \equiv \sum_{i=1}^{Y} \left(-\frac{1}{2}\Delta_i - \frac{Z}{|x_i|}\right) + \sum_{1\leq i<j\leq Y} \frac{1}{|x_i - x_j|} \quad .$$

In the case of (2) the constraint is that ψ belong to the set

$$S_s = \{\psi = \psi(x_1,x_2) | \psi(x_1,x_2) = u(x_1)u(x_2), \int u^2(x)dx = 1\} \ .$$

In the case of (1) the ψ is restricted to the set

$$S_{as} = \{\psi = \psi(x_1,\ldots,x_Y) | \psi = \frac{1}{\sqrt{Y!}} \begin{vmatrix} u_1(x_1)\ldots u_Y(x_1) \\ \\ u_1(x_Y)\ldots u_Y(x_Y) \end{vmatrix} \ ,$$

$$\int u_i(x)u_j(x)dx = \delta_{ij}\} \quad .$$

The critical points of $(\psi,H\psi)$ under the constraint $\|\psi\| = 1$ and ψ antisymmetric in its vector variables are the solutions of the Schrödinger equation which represent bound states for the atom. The reason to use (1) and (2) instead is that for radially symmetric functions there are iteration procedures for these equations which are much easier to handle numerically than the Schrödinger equation itself. The hope is then, that the solutions of (1) or (2) - if they exist - are reasonable approximations to the true bound states of the atom. The existence question for (1) and (2) has not been considered rigorously until recently. Before stating some results which have been obtained we want to make a simple transformation of the equations which is well known in the literature on Schrödinger operators. Performing a dilation of the independent variable $(x \to tx)$ there is a unitary transformation of $L^2(\mathbb{R}^3)$ generated by it: $u(x) \to t^{3/2}u(tx)$. This transforms (2) into

$$\left(-\frac{1}{2}\Delta - \frac{1}{|x|}\right)u(x) + \rho u(x)\int \frac{u^2(y)}{|x-y|}dy = \lambda' u(x)$$

$$\int u^2(x)dx = 1$$

and similar for (1). Thus $\rho = \frac{1}{Z}$ becomes a parameter governing the size of the third order term.

The first rigorous result for (2) was the author's paper [1] where it was shown that for all $Z > \sqrt{2}$ there exists a unique positive solution. Later Wolkowisky showed in [4] that there exists a $Z_o > 0$ such that (2) does not have solutions for $Z < Z_o$. The main subject of Wolkowisky's paper [4] is the study of a system of equations related to (1) but with a simpler structure of the nonlinearity. Finally Stuart showed in [3] that (2) has at least countably many solutions with definite nodal properties for all $Z > \sqrt{2}$. His result overlaps that of Wolkowisky but is achieved by entirely different methods. We now turn to (1) which is treated in [2]. We shall state here some of the results and give some indication by what methods they are deduced.

The first observation concerning (1) is that it has the symmetry group $O(Y)$. Let C be an orthogonal matrix then the transformation $\bar{u}'(x) = C\cdot\bar{u}(x)$ transforms (1) into a system of the same form, $\bar{\Lambda}$ being replaced by $C\bar{\Lambda}C^t$. The side conditions are obviously invariant under this transformation. It also follows from this remark that one can always find a C such that the transformed $\bar{\Lambda}$ becomes a diagonal matrix. The presence of this symmetry group implies that all solutions of (1) are degenerate. This creates troubles when one tries to use the topological degree in order to establish the existence of continua of solutions for different Z .

Multiplying the i-th equation by u_j , the j-th equation by u_i and integrating over $\mathbb{R}^3$ the difference of both terms gives $(\lambda_i-\lambda_j)(u_i,u_j)$. Thus if $\lambda_i \neq \lambda_j$ for $i \neq j$ then the corresponding u's are necessarily orthogonal. This leads to the following strategy: we drop all of the constraints except $\int u_i^2(x)dx = 1$ for $i = 1,\dots,Y$. Then we establish the existence of solutions for this system of equations and try to show après coup that for certain of these solutions $\lambda_i \neq \lambda_j$ for $i \neq j$.

In order to apply the topological degree we want to put the system in the form $(id + F)$ with a compact map F . For this purpose we introduce the following matrix of multiplication operators:

$$M(\bar{u}) = \begin{pmatrix} \sum_{i\neq 1}\int \frac{u_i^2(y)}{|x-y|}dy, & -\int\frac{u_1(y)u_2(y)}{|x-y|}dy, & \ldots, & -\int\frac{u_1(y)u_Y(y)}{|x-y|}dy \\ -\int\frac{u_Y(y)u_1(y)}{|x-y|}dy, & \ldots, & -\int\frac{u_Y(y)u_{Y-1}(y)}{|x-y|}dy, & \sum_{i\neq Y}\int\frac{u_i^2(y)}{|x-y|}dy \end{pmatrix}$$

We recall that we suppose $\bar{\Lambda}$ to be diagonal. If we further drop all side conditions except $\int u_i^2(x)dx = 1$ then we get

$$(-\tfrac{1}{2}\Delta+\tfrac{1}{Z}M(\bar{u})-\bar{\Lambda})\bar{u} = \frac{1}{|x|}\bar{u}$$

$$\int u_i^2(x)dx = 1 \qquad i = 1,\ldots,Y .$$

Here $-\frac{1}{2}\Delta$ stands for the matrix operator which has $-\frac{1}{2}\Delta$ in the diagonal and zero in all off-diagonal terms. This operator is a self-adjoint operator on $\sum_1^Y \oplus L^2(\mathbb{R}^3)$ with domain $\sum_1^Y \oplus L_2^2(\mathbb{R}^3)$ and spectrum $[0,\infty)$. For $u_i \in L_1^2(\mathbb{R}^3)$ $M(\bar{u})$ is a small perturbation of $-\frac{1}{2}\Delta$ in the sense of Kato and the spectrum of $-\frac{1}{2}\Delta + \frac{1}{Z}M(\bar{u})$ is contained in $[0,\infty)$ because $M(\bar{u})$ is non-negative. Thus, if $\bar{\Lambda} \leqslant -\delta$ then the operator $-\frac{1}{2}\Delta + \frac{1}{Z}M(\bar{u}) - \bar{\Lambda}$ is invertible, its inverse being a bounded operator from $\sum_1^Y \oplus L^2(\mathbb{R}^3)$ to $\sum_1^Y \oplus L_2^2(\mathbb{R}^3)$ with bound $1/\delta$. Therefore we can rewrite the above system in the following final form:

$$(3)\qquad \begin{aligned} \bar{u} &= \left(-\tfrac{1}{2} + \tfrac{1}{Z}M(\bar{u})-\bar{\Lambda}\right)^{-1}\tfrac{1}{|x|}\bar{u} \\ \int u_i^2(x)dx &= 1 \qquad i = 1,\dots,Y\ . \end{aligned}$$

We define F_ρ as a function from $\sum_1^Y \oplus L_2^2(\mathbb{R}^3) \times \prod_1^Y \mathbb{R}$ to itself by

$$F_\rho(\bar{u},\bar{\Lambda}) = \left(\left(-\tfrac{1}{2}\Delta+\rho M(\bar{u})-\bar{\Lambda}\right)^{-1}\tfrac{1}{|x|}\bar{u},\lambda_i-\int u_i^2(x)dx+1\right)\ .$$

Then the equations (3) are equivalent to

$$(4)\qquad (\bar{u},\bar{\Lambda}) - F_\rho(\bar{u},\bar{\Lambda}) = 0\ .$$

It can be shown that F_ρ maps any bounded set of $\sum_1^Y L_2^2(\mathbb{R}^3) \times \prod_1^Y \mathbb{R}^1$ with $\bar{\Lambda} \leqslant -\delta$ into a precompact set. Further F_ρ is Frechet differentiable with respect to u , $\bar{\Lambda}$, ρ . For $\rho = 0$ the solutions of (3) are explicitly known and it can be checked that the Frechet derivative at these points is invertible if one restricts F_ρ to the space of radially symmetric functions. Thus, the fixed point index is $+1$ or -1 . It follows from wellknown properties of the topological degree that all solutions $(u,\bar{\Lambda},0)$ are starting points of continua of solutions of (4). These continua are either unbounded in the $(\bar{u},\bar{\Lambda},\rho)$ space or they connect two solutions for $\rho = 0$ or at least one of the λ_i tends to zero.

Looking at (1) we see that the system has the form

$$L_i u_i = \lambda_i u_i$$

where the linear operator L_i has the form

$$L_i = -\frac{1}{2}\Delta - \frac{1}{|x|} + \rho\int\frac{\sum_{j\neq i} u_j^2(y)}{|x-y|}dy - \rho\int\frac{\sum_{j\neq i} u_j(x)u_j(y)\cdot}{|x-y|}dy .$$

The point indicates that the function L_i acts upon has to be put under the integral sign. The last term in L_i is a non-negative integral operator. On the space of radial functions the following estimates are therefore valid

$$-\frac{1}{2}\Delta - \frac{1}{|x|} < L_i < -\frac{1}{2}\Delta - \frac{1-\rho(Z'-1)}{|x|} .$$

Along a continuum of solutions the operator L_i changes continuously starting with $-\frac{1}{2}\Delta - \frac{1}{|x|}$ for $\rho = 0$. Let $\lambda_i^{(0)}$ be any set of Y distinct increasing eigenvalues of $-\frac{1}{2}\Delta - \frac{1}{|x|}$ and let n_i be the number of $\lambda_i^{(0)}$ in the natural ordering of the eigenvalues. If

$$-\frac{(1-\rho(Z'-1))^2}{n_i^2} \leqslant -\frac{1}{(n_i+1)^2} \quad \text{for all } i$$

then the continuum starting at $(u_i^{(0)},\lambda_i^{(0)},0)$ reaches the set $\{\rho = \frac{1}{Z}\}$. The reason is that λ_i stays the n_i-th eigenvalue of L_i . Thus it can neither come back to another eigenvalue at $\rho = 0$ nor can one of the λ_i reach zero.

This result is good only for $Z \gg Y$ and even then it gives only a finite number of bound states. For $Z = Y$ it gives nothing at all. If one knew that the operators L_i and the operator

$$-\frac{1}{2}\Delta - \frac{1}{|x|} + \rho\int\frac{\Sigma u_i^2(y)}{|x-y|}dy - \rho\int\frac{\Sigma u_i(x)u_i(y)\cdot}{|x-y|}dy$$

have only simple eigenvalues on the space of radial functions then one could do away with the above comparison argument and get the very strong and satisfactory result that from all solutions for $\rho = 0$ there are continua emanating which do not intersect and which do reach the set $\{\rho = \frac{1}{Z}\}$. But at the moment we see no way of proving that these operators have only simple eigenvalues, the trouble coming from the integral operator.

In order to get nevertheless results for $Z = Y$ we have to use variational methods. The essential technical condition to be fullfilled is condition C. For the variational procedure we consider the functional

$$f_\rho(\bar{u}) = \int\bar{u}(x)\cdot\left(-\frac{1}{2}\Delta-\frac{1}{|x|}\right)\bar{u}(x)dx + \rho\iint\frac{\bar{u}^2(x)\bar{u}^2(y)}{|x-y|}dxdy$$
$$- \rho\iint\frac{(\bar{u}(x)\cdot\bar{u}(y))^2}{|x-y|}dxdy$$

on the space $L_1^2 = \prod_1^Y L_1^2(\mathbb{R}^3)$ restricted to the set defined by

$\int u_i(x)u_j(x)dx = \delta_{ij}$. This set is a C^∞-manifold M in L_1^2 . The derivative of f_ρ with respect to $\bar{u}$ is a linear functional on the tangent plane to M . Using the natural pairing between L_1^2 and $L_{-1}^2 = \prod_1^Y L_{-1}^2(\mathbb{R}^3)$ we get the gradient as an element of L_{-1}^2 . We shall denote the gradient of f_ρ at $\bar{u}$ by $\bar{v}$.

$$\bar{v}(x) = \left[-\frac{1}{2}\Delta-\frac{1}{|x|}\right]\bar{u}(x) + \rho\bar{u}(x)\int\frac{\bar{u}^2(y)}{|x-y|}dy$$

$$(5) \qquad - \rho\int\frac{(\bar{u}(x)\cdot\bar{u}(y))\bar{u}(y)}{|x-y|}dy - \bar{\Lambda}\cdot\bar{u}(x) .$$

As we know it is permissible to assume $\bar{\Lambda}$ in diagonal form. We rewrite (5) in the same fashion as before:

$$\bar{u} = \left(-\frac{1}{2}\Delta+M(\bar{u})-\bar{\Lambda}\right)^{-1}\left[\frac{1}{|x|}\bar{u}+\bar{v}\right] .$$

We assume that $\bar{u}_i$ be a sequence in M with $\bar{\Lambda}_i \leqslant -\delta$, $f_\rho(\bar{u}_i) \leqslant$ const. and $\bar{v}_i \to 0$ in L_{-1}^2 . Then there is a subsequence $\bar{u}_{i'}$ such that $\bar{u}_{i'} \to \bar{u}$ in L_1^2 and $\bar{u}$ is a critical point of f_ρ on M . The argument is similar to the one proving compactness of F_ρ only now all the maps have to be lifted to maps between L_1^2 and L_{-1}^2 . The remaining problem is to relate the values of the sequence to the λ_i . Let S_Y be the sum of the first $Y - 1$ eigenvalues of the Schrödinger operator for hydrogenlike atoms of charge Z . Then all the values of f_ρ smaller than S_Y

are normal that is condition C is satisfied for these values. For the proof we remark that the fourth order term in f_ρ is positive. Thus, if $f < S_Y$ then $(\bar{u},(-\frac{1}{2}\Delta-\frac{1}{|x|})\bar{u}) \leqslant S_Y$ and this in turn implies that $(u_i,(-\frac{1}{2}\Delta-\frac{1}{|x|})u_i) < 0$. By multiplying the i-th equation by u_i and integrating we can express the λ_i in terms of the u_i..

$$\lambda_i = (u_i,(-\tfrac{1}{2}\Delta-\tfrac{1}{|x|})u_i) + \rho\sum_{j\neq i}\int\frac{u_i^2(x)u_j^2(y)}{|x-y|}dxdy$$

$$- \rho\sum_{j\neq i}\int\frac{u_i(x)u_j(x)u_i(y)u_j(y)}{|x-y|}dxdy$$

$$\leqslant (u_i,(-\tfrac{1}{2}\Delta-\tfrac{1-\rho(Z'-1)}{|x|})u_i) < 0 \quad \text{for} \quad \rho^{-1} > Z' - 1 .$$

The last inequality follows from $(u_i,(-\frac{1}{2}\Delta-\frac{1}{|x|})u_i) < 0$ by a scaling transformation.

This proves that values smaller than S_Y are normal and it also follows from the above argument that there is a set of infinite category (after identification of opposite points on M) where $f < S_Y$. It follows that there are at least countably many critical values for f. This finishes the existence proof for all $Z \geqslant Y$.

REFERENCES

[1] Reeken, M.: General Theorem on Bifurcation and Its Application to the Hartree Equation of the Helium Atom. Journal of Mathematical Physics Vol. 11, No 8, 2505-2512 (1970).

[2] Reeken, M.: Existence of Solutions for the Hartree and Hartree-Fock Equations. To appear in the Proceedings of the Royal Society of Edinburgh.

[3] Stuart, C.A.: Existence Theory for the Hartree Equation, Archive for Rational Mechanics and Analysis, Vol. 51, No. 1, 60-69 (1973).

[4] Wolkowisky, J.H.: Existence of Solutions of the Hartree Equation for N Electrons. Indiana Math. Journal, Vol. 22, No. 6, 551-568 (1972).

CENTRO INTERNAZIONALE MATEMATICO ESTIVO

(C. I. M. E.)

POSITIVE SOLUTIONS OF NONLINEAR EIGENVALUE PROBLEMS

R. E. L. TURNER

Corso tenuto a Veranna dal 16 al 25 giugno 1974

POSITIVE SOLUTIONS OF NONLINEAR EIGENVALUE PROBLEMS

by

R. E. L. Turner
(University of Wisconsin, Madison)

1. A Nonlinear Krein-Rutman Theorem

Let X be a real Banach space and let K be a cone in X; i.e., K is i) closed, ii) convex, and iii) $u \in K$ and $\alpha \geq 0$ implies $\alpha u \in K$. We further assume that K is proper: $K \cap (-K) = \{0\}$ and that each vector $u \in X$ can be represented as $u = u_1 - u_2$ with u_1 and u_2 in K. Consider a continuous map $F = F(\lambda, u)$ from $[0, \infty) \times K$ into K satisfying $F(\lambda, 0) = 0$ for all $\lambda \geq 0$. One says F is compact if F maps each bounded set in $[0, \infty) \times K$ into a precompact set in K. We say F has a derivative at $u = 0$, denoted $dF(0)$, if $dF(0)$ is a linear map satisfying $F(\lambda, u) - F(\lambda, 0) - dF(0)u = o(\|u\|)$ for $u \in K$, uniformly in λ.

The problem

$$u = \lambda F(\lambda, u) \tag{1.1}$$

will have the trivial solution $(\lambda, 0)$ for any $\lambda \geq 0$. Letting $R^+ = [0, \infty)$ we define a set of nontrivial solutions

$$S' = \{(\lambda, u) \in R^+ \times (K - \{0\}) \mid u = \lambda F(\lambda, u)\} \tag{1.2}$$

and let $S = \overline{S'}$ where the bar denotes the closure. The set S may contain trivial solutions, but only those that are limits of nontrivial solutions; i.e., only the bifurcation points of (1.1) (cf. [1], p. 207). In each of the results below we establish the existence of a continuum (i. e. a closed, connected set) of solutions of an equation of type (1. 1). We will see that varying assumptions on F will enable us to give information about the location of the continuum in $R^+ \times K$. The first result is a nonlinear version of the classical Krein-Rutman theorem for linear operators (cf. [2], Theorem 6. 1).

Theorem 1. 1. Let $F = F(\lambda, u)$ be a compact map of $R^+ \times K$ into K, satisfying $F(\lambda, 0) = 0$ for $\lambda \geq 0$. Suppose F has a derivative $dF(0)$ at $u = 0$ and that $dF(0)$ has a spectral radius $\lambda_1^{-1} > 0$. Then S contains an unbounded continuum C with $(\lambda_1, 0) \in C$.

The proof of Theorem 1. 1, which depends upon the use of transversality arguments, can be found in [3] (cf. also [4], section 3). E. N. Dancer [5] has independently proved Theorem 1. 1 using methods based on degree theory.

We next give an application of Theorem 1. 1 to a system of elliptic partial differential equations. Let Ω be a bounded domain in R^n and assume its boundary $\partial\Omega$ is of class $C^{2+\alpha}$ (cf. [6], p. 6). We use $C^k(\Omega)$ to denote the Banach space of real valued functions which have k continuous derivatives on Ω and use $C^k(\bar{\Omega})$ to denote the set of functions in $C^k(\Omega)$ which are continuously extendable to $\bar{\Omega}$ in the sense of [6], page 4. We use $C^{k+\alpha}(\bar{\Omega})$ to denote the space of those functions v in $C^k(\bar{\Omega})$ for which each mixed k^{th} derivative of v satisfies a Hölder condition with exponent α, $0 < \alpha < 1$. We denote the standard norms in C^k and $C^{k+\alpha}$

by $\| \ \|_k$ and $\| \ \|_{k+\alpha}$, respectively. The symbols $\overset{\circ}{C}{}^k(\Omega)$ and $\overset{\circ}{C}{}^{k+\alpha}(\bar{\Omega})$ will denote the subspaces of $C^k(\bar{\Omega})$ and $C^{k+\alpha}(\bar{\Omega})$, respectively, containing functions which vanish on $\partial\Omega$. A subscript m, e.g. $C^k_m(\bar{\Omega})$, denotes the space of m-tuples of functions $v = (v_1, v_2, \ldots, v_m)$ such that each component v_i is in the unsubscripted space. With addition and scalar multiplication defined on components, $C^k_m(\Omega)$, $\overset{\circ}{C}{}^k_m(\bar{\Omega})$, $C^{k+\alpha}_m(\bar{\Omega})$, and $\overset{\circ}{C}{}^{k+\alpha}_m(\bar{\Omega})$ become real Banach spaces and we write

$$\|v\|_k = \sup_{1 \le i \le m} \|v_i\|_k \tag{1.3}$$

or a similar expression with subscript $k + \alpha$ to denote the corresponding norm. In the application we use the cone

$$K = \{v = (v_1, \ldots, v_m) \in \overset{\circ}{C}{}^{1+\alpha}_m(\bar{\Omega}) \mid u_i \ge 0; \ i = 1, 2, \ldots m\} \tag{1.4}$$

For $v = (v_1, \ldots v_n)$ in $C^1_m(\bar{\Omega})$, we let ∇v stand for the collection of gradients $(\nabla v_1, \ldots, \nabla v_m)$. Then for each $x \in \bar{\Omega}$, $v(x)$ is a point $p \in R^m$ and ∇v is a point $q \in R^{mn}$. If u is a vector in $C^{1+\alpha}_m(\bar{\Omega})$ and w is a function in $C^2(\Omega)$, we let L^k_u $(k = 1, 2, \ldots, m)$ be the elliptic expression defined by

$$L^k_u w = - \sum_{i,j=1}^{n} a^k_{ij}(x, u, \nabla u) \frac{\partial^2 w}{\partial x_i \partial x_j} + \sum_{i=1}^{n} b^k_i(x, u, \nabla u) \frac{\partial w}{\partial x_i} + c^k(x, u, \nabla u) w \tag{1.5}$$

where a^k_{ij}, b^k_i, and $c^k \ge 0$ are C^1 functions on $\bar{\Omega} \times R^m \times R^{mn}$ and a^k_{ij} is positive definite:

$$\sum_{i,j=1}^{n} a^k_{ij}(x, p, q)\xi_i\xi_j \ge e \sum_{i=1}^{n} \xi_i^2; \ (\xi_1 \ldots \xi_n) \in R^n \tag{1.6}$$

for some $e > 0$ and $1 \le k \le m$. Further, for $1 \le k \le m$, we let a^k be a strictly positive function in $C^{r}(\bar{\Omega})$ and let $h^k = h^k(x, p, q)$ be a nonnegative C^1 function on $\bar{\Omega} \times R^m \times R^{mn}$. It is assumed that h^k is $o(|p| + |q|)$ as $|p| + |q| \to 0$, where $|p|$ and $|q|$ are the Euclidean norms of p and q.

Theorem 1.2. For $1 \leq k \leq m$ and $1 \leq i, j \leq n$ let the quantities a^k_{ij}, b^k_i, c^k, L^k_u, a^k, and h^k be as defined above. Let λ^k be the smallest nonnegative eigenvalue of

$$L^k_0 w = \lambda a^k w \tag{1.7}$$

with $w \in C^2(\Omega) \cap \overset{\circ}{C}{}^0(\bar{\Omega})$, and let $\lambda_1 = \min(\lambda^1, \ldots, \lambda^m)$. Consider the system

$$- \sum_{i,j=1}^{n} a^k_{ij}(x, u, \nabla u) \frac{\partial^2 u_k}{\partial x_i \partial x_j} + \sum_{i=1}^{n} b^k_i(x, u, \nabla u) \frac{\partial u_k}{\partial x_i}$$

$$+ c^k(x, u, \nabla u) u = \lambda(a^k(x) u_k + h^k(x, u, \nabla u)); \quad k = 1, 2, \ldots, m, \tag{1.8}$$

with $u = (u_1, \ldots u_m)$ and let S be the closure in $R^+ \times K$ of the nontrivial solutions of (1.8) in $R^+ \times (C^2_m(\Omega) \cap K)$. Then $\lambda_1 > 0$ and S contains an unbounded continuum C with $(\lambda_1, 0)$ in C.

Proof. Letting $\mathcal{L}_u$ stand for the diagonal matrix of operators with $L^1_u, \ldots, L^m_u$ on the diagonal; introducing a diagonal matrix a with entries $a^1, \ldots, a^n$; and a vector $h = (h^1, \ldots, h^m)$, one can write the problem (1.8) in the form

$$\mathcal{L}_u u = \lambda(au + h(x, u, \nabla u) \tag{1.9}$$

or

$$u = \lambda F(u) \tag{1.10}$$

with $F(u) = \mathcal{L}_u^{-1}(au + h)$. One shows that F satisfies the hypotheses of Theorem 1.1 using the fact that $\mathcal{L}_u^{-1}$ maps nonnegative functions to nonnegative functions. The existence of the desired continuum then follows from Theorem 1.1. See [3] for details.

We will call solutions in C "positive."

In the case $m = 1$, the last result was obtained by Rabinowitz [7], using the simplicity of λ_1 as an eigenvalue of $L_0^1 w = a^1 w$. In our application λ_1 need not be a simple eigenvalue of $\mathcal{L}_0 u = \lambda au$. In [7], the assumption $h^1 = h \geq 0$ was not made, but it was assumed that $h(x, p, q) = o(|p|)$ as $|p| \to 0$. A continuum for this case can be obtained by writing h as $h^+ - h^-$ where $h^\pm \geq 0$ and $h^\pm$ is $o(|p|)$ as $|p| \to 0$. Then with $u = u_1$, $L_u u = \lambda(a^1 u + h)$ is equivalent to $(L_u + \lambda \frac{h^-}{u})u = \lambda(au + h^+)$, a problem which can be treated in much the same way as (1.8) for $m = 1$.

2. Sublinear Operators

We say an operator $F = F(u)$ is homogeneous of degree s if $F(tu) = t^s F(u)$ for real $t \geq 0$. While we don't give a precise definition of "sublinear," we have in mind as examples the operators which are homogeneous of degree s with $0 < s \leq 1$. In the case that $s \geq 1$, we would call such an operator "superlinear." In this section we will be interested in sublinear behavior near $u = 0$.

We use the standard partial order induced by a cone $K \subset X$ and write $x \leq y$ or $y \geq x$ if and only if $y - x \in K$. A map B from K into K is called <u>monotone</u> if and only if $Bx \geq By$ whenever $x \geq y$.

We next give a theorem and corollary which illustrate general assumptions under which one can obtain the existence of solutions of eigenvalue problems for mappings in cones. The aim here is to contrast the sublinear and superlinear cases (cf. section 3). For convenience we suppose that F is independent of λ and maintain the definitions of S' (cf. (1.2)) and S. Theorem 2.1 and Corollary 2.2 extend results of Krasnoselskii ([1], section 5.3).

Theorem 2.1. Let F be a compact, continuous map of a cone K into itself, and assume $F(0) = 0$. Suppose B is a monotone map, homogeneous of degree 1, and that $Bu_0 \geq \delta u_0$ for some $u_0 \in K - \{0\}$ and some real $\delta > 0$. If $Fu \geq Bu$ for $u \in K$ and $0 < \|u\| < \rho$, then

$$u = \lambda F(u),$$

the set S contains an unbounded continuum C such that

$C \cap \{[0, \delta^{-1}] \times \{0\}\} \neq \emptyset$ and $C \cap \{(\delta^{-1}, \infty) \times \{0\}\} = \emptyset$.

Proof. As an easy corollary of Lemma 1.4 of [3], one sees that there is a linear functional ℓ, with $\ell \geq 0$ on K, which takes positive values on $F(K) - \{0\}$ and which satisfies $\langle \ell, u_0 \rangle = 1$. We let B_0 be the rank one map defined by $B_0 u = \delta\langle \ell, u\rangle u_0$. Let $\eta = \eta(t)$ be a C^∞ function from $[0, \infty)$ to the interval $[0, 1]$ such that $\eta = 0$ for $t \leq 1$ and $\eta = 1$ for $t \geq 2$. Consider the map

$$F_n u = (1 + \frac{1}{n} - \eta(n\|u\|))B_0 u + \eta(n\|u\|)Fu \tag{2.1}$$

for $n = 1, 2, 3, \ldots$. The map F_n is compact, continuous and, as it is equal to $(1 + \frac{1}{n})B_0$ for u near $u = 0$, it has a derivative $(1 + \frac{1}{n})B_0$ at

$u = 0$. It follows from Theorem 1, 1 that there is an unbounded continuum C_n in the solution set $S \subset R^+ \times K$ for $u = \lambda F_n u$ and that $(\delta^{-1}(1 + \frac{1}{n})^{-1}, 0)$ is in C_n. Since B_0 has rank one, no other trivial solution can be in C_n. We define

$$\limsup C_n = \{(\lambda, u) \in R^+ \times K \mid \exists n_1 < n_2 < \ldots ; (\lambda_{n_k}, u_{n_k}) \in C_{n_k};$$

$$\tag{2.2}$$

The set $\tilde{C} = \limsup C_n$ will be an unbounded continuum (cf. [3]) and will consist of solutions of $u = \lambda F u$ inasmuch as F_n approaches F uniformly on K as $n \to \infty$. We next determine where $\tilde{C}$ meets the set of trivial solutions.

Suppose $Fu \geq Bu$ for $u \in K$ and $0 < \|u\| < \rho$. Let (λ_n, u_n) be in C_n with $0 < \|u_n\| < \rho$. Since $u_n \neq 0$, $\lambda_n \neq 0$ and

$$\lambda_n^{-1} u_n = (1 + \frac{1}{n} - \eta(n\|u_n\|)B_0 u_n + \eta(n\|u_n\|)Fu_n$$

$$\geq (1 + \frac{1}{n} - \eta)B_0 u_n + \eta B u_n \tag{2.3}$$

One shows easily, using (2.3), that $\langle \ell, u_n \rangle > 0$. As $\lambda_n^{-1} u_n \geq n^{-1}\delta\langle \ell, u_n\rangle u_0$, it follows that $u_n \geq t_n u_0$ for some $t_n > 0$. We suppose t_n to be the largest such number, which will exist, as K is closed. Then

$$\begin{aligned} \lambda_n^{-1} u_n &\geq (1 + \frac{1}{n} - \eta)B_0(t_n u_0) + \eta B(t_n u_0) \\ &\geq (1 + \frac{1}{n} - \eta)t_n \delta u_0 + \eta t_n u_0 \\ &\geq (1 + \frac{1}{n})t_n \delta u_0 . \end{aligned}$$

From the characterization of t_n, it follows that $\lambda_n(1+\frac{1}{n})\delta t_n \leq t_n$ or that $\lambda_n \leq \delta^{-1}$. Thus $\tilde{C} \cap \{(\delta^{-1}, \infty) \times \{0\}\} = \emptyset$.

While one can easily show that $\tilde{C} \cap \{[0, \delta^{-1}] \times \{0\}\} \neq \emptyset$, the continuum $\tilde{C}$ may contain trivial solutions of $u = \lambda F(u)$ which are not in the solution set S. In the next paragraph we obtain a subcontinuum of $\tilde{C}$ which does lie in S.

For each integer $m > \rho^{-1}$ let $S^m = \{(\lambda, u) \in R^+ \times K \mid 0 \leq \lambda \leq \delta^{-1}, 0 \leq \|u\| < \frac{1}{m}\}$ and let ∂S^m be its boundary relative to $R^+ \times K$. Since $\tilde{C}$ is an unbounded continuum which contains $(\delta^{-1}, 0)$ and which contains no points (λ, u) with $\lambda > \delta^{-1}$ and $0 \leq \|u\| < \rho$, it follows that $\partial S^m \cap \tilde{C} \neq \emptyset$ and further that $\partial S^m \cap \tilde{C}$ contains a pair (λ, u) with $\|u\| = \frac{1}{m}$. We add a point at "∞" to each of R^+ and K using the standard stereographic projection and induced metric. Since F maps bounded sets in $R^+ \times K$ to precompact sets in K, one can show the closure in $(R^+ + \infty) \times (K + \infty)$ of the set of solutions of $u = \lambda Fu$ in $R^+ \times K$, is a compact set. Then, using properties of continua (cf. [8], p. 12), one sees that for some point $(\lambda^m, u^m) \in \tilde{C}$ with $\|u^m\| = \frac{1}{m}$, the component C^m of (λ^m, u^m) in $\tilde{C} - S^m$ will contain a point of $[(R^+ + \infty) \times (\infty)] \cup [\infty \times (K + \infty)]$; i.e., the component C^m will be unbounded in $R^+ \times K$. The sequence (λ^m, u^m) will have a limit point $(\lambda_0, 0)$ with $0 \leq \lambda_0 \leq \delta^{-1}$ and we assume without loss of generality that (λ^m, u^m) converges to $(\lambda_0, 0)$ as $n \to \infty$. Let L^m be the closed straight line segment from $(\lambda_0, 0)$ to (λ^m, u^m) and define: $\tilde{C}^m = C^m \cup L^m$. The set

$C = \limsup \tilde{C}^m$, as $m \to \infty$, is an unbounded continuum which is contained in $\tilde{C}$ and which contains $(\lambda_0, 0)$. Each point of C of the form $(\lambda, 0)$ is a limit point of nontrivial solutions in $\tilde{C}$ and hence $C \subset S$. The continuum C then satisfies the requirements in the conclusion of the theorem.

Theorem 2. 2. If B, in theorem 2. 1, is assumed to be homogeneous of degree s, $0 < s < 1$, then there is an unbounded continuum C of solutions of $u = \lambda F(u)$. The point $(0, 0) \in C$ and there is a constant $M > 0$ such that $(\lambda, u) \in C$ implies $\lambda \leq M\delta^{-1} \|u\|^{1-s}$ for $0 \leq \|u\| < \rho$.

Proof. We proceed as in the previous proof, but replace (2. 1) by

$$F_n u = (1 + \frac{1}{n} - \eta) n \langle \ell, u \rangle u_0 + \eta F u$$

where $\eta = \eta(n\|u\|)$. We obtain an unbounded continuum C_n of solutions of $u = \lambda F_n u$. If $(\lambda_n, u_n) \in C_n$ with $0 < \|u_n\| < \rho$, then, as before there is a maximal $t_n > 0$ such that $t_n u_0 \leq u_n$. Since $t_n u_0 \leq u_n$, $\langle \ell, t_n u_0 \rangle \leq \langle \ell, u_n \rangle$ or $t_n \leq \|\ell\| \cdot \|u_n\|$. Hence

$$\begin{aligned} \lambda_n^{-1} u_n &= (1 + \frac{1}{n} - \eta) n \langle \ell, u_n \rangle u_0 + \eta F u_n \\ &\geq (1 + \frac{1}{n} - \eta) n \langle \ell, t_n u_0 \rangle u_0 + \eta B(t_n u_0) \\ &\geq (1 + \frac{1}{n} - \eta) n t_n u_0 + \eta t_n^s B u_0 \\ &\geq (1 + \frac{1}{n} - \eta) n t_n u_0 + \eta t_n^s \delta u_0 \\ &\geq [(1 + \frac{1}{n} - \eta) n + \eta \delta (\|\ell\| \cdot \|u_n\|)^{s-1}] t_n u_0 \end{aligned}$$

where $\eta = \eta(n\|u_n\|)$. It follows from the characterization of t_n that

$$\lambda_n \leq [(1 + \frac{1}{n} - \eta) n + \eta \delta (\|\ell\| \cdot \|u_n\|)^{s-1}]^{-1}.$$

It follows that $\lambda_n \leq \max(n^{-1}, \delta^{-1}\|\ell\|^{1-s}\|u_n\|^{1-s})$ for all n and that $\lambda_n \leq \delta^{-1}\|\ell\|^{1-s}\|u_n\|^{1-s}$ if $n\|u_n\| \geq 2$, in which case $\eta = 1$. If $(\lambda_n, 0) \in C_n$, then $\lambda_n = (n+1)^{-1}$. The continuum we seek is $C = \lim\sup C_n$. It contains $(0, 0)$, is unbounded, and satisfies the desired inequalities, as one sees by passing to the limit as $n \to \infty$ in the inequalities just established.

The next theorem was proved by H. Kuiper [9] for the case in which the map F is independent of λ. We allow dependence on λ, as we will need it for a later application. The proof in [9] uses results of Kransnoselskii ([1], p. 161), based on degree theory. We give a proof based on Theorem 1.1.

Theorem 2.3. Let $F = F(\lambda, u)$ be a compact, continuous map of $R^+ \times K$ into K and suppose $F(\lambda, 0) \neq 0$ for all $\lambda \geq 0$. Then the set of solutions of $u = \lambda F(\lambda, u)$ in $R^+ \times K$ contains an unbounded continuum C with $(0, 0) \in C$.

Proof. As before, we alter F near $u = 0$ to make it linear. Let ℓ be a linear functional which is nonnegative on K, positive on $F(\lambda, 0)$ for $\lambda \geq 0$, and which satisfies $\langle \ell, F(0, 0)\rangle = 1$. The existence of ℓ is an easy corollary of [3], Lemma 1.4. Let

$$F_n(\lambda, u) = (1 - \eta(n\|u\|))n\langle \ell, u\rangle F(0, 0) + \eta(n\|u\|)F(\lambda, u).$$

Then we can apply Theorem 1.1 to $u = \lambda F_n(\lambda, u)$ to obtain a continuum C_n of solutions emanating from $(\frac{1}{n}, 0)$. The set $C = \lim\sup C_n$ is an unbounded continuum containing $(0, 0)$ and one verifies easily that any $(\lambda, u) \in C$ with

$u \neq 0$ satisfies $u = \lambda F(\lambda, u)$, since F_n converges to F, uniformly on any set $\{(\lambda, u) \mid \|u\| \geq \epsilon > 0\}$. If we show that $(\lambda_{n_k}, u_{n_k}) \in C_{n_k}$ and $\lim_{k \to \infty} (\lambda_{n_k}, u_{n_k}) = (\bar{\lambda}, 0)$ implies $\bar{\lambda} = 0$, then the conclusion of the theorem will follow. Suppose that $(\bar{\lambda}, 0)$ is such a limit and that $\bar{\lambda} > 0$. Then with $\eta_k = \eta(n_k \|u_{n_k}\|)$,

$$u_{n_k} = \lambda_{n_k}[(1 - \eta_k)n_k\langle \ell, u_{n_k}\rangle F(0, 0) + \eta_k F(\lambda_{n_k}, u_{n_k})] \tag{2.4}$$

and

$$\langle \ell, n_k u_{n_k}\rangle = n_k \lambda_{n_k}[(1 - \eta_k)\langle \ell, n_k u_{n_k}\rangle + \eta_k\langle \ell, F(\lambda_{n_k}, u_{n_k})\rangle] \tag{2.5}$$

If $\eta_k \geq \frac{1}{2}$ for an infinite subsequence of $\{n_k\}$, then from (2.4) one obtains $0 \geq \bar{\lambda} \cdot \frac{1}{2} F(0, 0)$, an impossibility. Otherwise, $\eta_k \leq \frac{1}{2}$ for an infinite subsequence of $\{n_k\}$. If $\eta_k = 0$ in (2.4), then since $u_{n_k} \neq 0$, $\langle \ell, u_{n_k}\rangle > 0$. if $\eta_k > 0$, then since $\langle \ell, F(\lambda_{n_k}, u_{n_k})\rangle > 0$ for large k, it follows, by applying ℓ to both sides of (2.4), that $\langle \ell, u_{n_k}\rangle > 0$. Having $\langle \ell, u_{n_k}\rangle > 0$ and $\eta_k \leq \frac{1}{2}$, one sees that (2.5) cannot be satisfied for n_k large.

While theorems 2.1, 2.2, and 2.3 can be applied to a wide class of nonlinear elliptic differential equations and integral equations, one can obtain stronger results either by proving abstract theorems tailored to applications or by working more directly with the particular class of equations. We choose to do the latter in Theorem 2.5. For that purpose we need the following lemma. We refer the reader to [6] for definitions of the Sobolev spaces $\overset{\circ}{W}{}^k_p(\Omega)$ used in the proof.

Lemma 2.4. For $u \in C^{1+\alpha}(\bar{\Omega})$, $u \geq 0$, and $w \in C^2(\Omega)$, let $L_u w$ be the elliptic expression defined by (1.5) for $k = m = 1$ and suppose a^1_{ij}, b^1_i, and c^1 are $C^{1+\alpha}$ functions of their arguments. Let a be $C^1(\bar{\Omega})$ and satisfy $a(x) \geq \sigma > 0$. Then there is a positive constant d such that for $\|u\|_0 < d$, the smallest eigenvalue λ_a of the problem $L_u w = a(x)w$, $w \in C^2(\Omega) \cap \mathring{C}^0(\bar{\Omega})$, satisfies

$$\lambda_a \leq \frac{2\lambda_1}{\sigma} \tag{2.6}$$

where λ_1 is the smallest eigenvalue of the problem $L_0 w = \lambda w$.

Proof. Since a^1_{ij} is in $C^{1+\alpha}(\bar{\Omega})$, the expression $L_0 v$ can be put in divergence form. As such, there is a maximum principle and uniqueness for a weak solution $v \in \mathring{W}^1_2(\Omega)$ of $L_0 v = f$, with $f \in L^2(\Omega)$ (cf. [10]). The uniqueness together with results from ([11], chap. 5) imply that $L_0 v = f$ has a solution $v \in \mathring{W}^2_2(\Omega)$ and that there is an $M > 0$, independent of f, such that $\|u\|_{2,2} \leq M\|f\|_{L^2}$ where $\| \; \|_{2,p}$ is the norm in $\mathring{W}^2_p(\Omega)$. If we let $L_u - L_0 = L_1$, then $L_1 L_0^{-1}$ is bounded as a map of $L^2(\Omega)$ into itself, since the range of L_0^{-1} is in $\mathring{W}^2_2(\Omega)$ and L_1 involves derivatives of order at most two. Moreover, the derivatives are multiplied by functions which can be made small in $C^0(\bar{\Omega})$ by requiring that $\|u\|_0$ be small. Thus, given $\epsilon > 0$, there is a $d > 0$ such that $\|L_1 L_0^{-1}\| < \epsilon$ provided $\|u\|_0 < d$.

While the lemma concerns the lowest eigenvalue of L_u as a map with domain $C^2(\Omega) \cap \mathring{C}^0(\bar{\Omega})$, L_u as a map with domain $\mathring{W}^2_2(\Omega)$ has the same eigenvalues, as can be seen using the regularity theory for elliptic equations ([11], [12]). It is known that λ_1, the lowest eigenvalue of L_0, is simple and corresponds to a positive eigenfunction w; i.e., $w > 0$ in Ω (cf. [3], section 4). Since

$$\begin{aligned} L_u^{-1} &= (L_0 + L_1)^{-1} \\ &= ((I + L_1 L_0^{-1})L_0)^{-1} \\ &= L_0^{-1} + L_0^{-1} L_1 L_0^{-1} (I + L_1 L_0^{-1})^{-1}, \end{aligned}$$

it follows from standard perturbation theory for linear operators ([13], chap. 7) that the largest eigenvalue of L_u^{-1} will be within any desired distance of λ_1^{-1}, provided $\| L_1 L_0^{-1} \|$ is sufficiently small. Thus there is a $d > 0$ such that when $\| u \|_0 < d$, the smallest eigenvalue of L_u is at most $2\lambda_1$.

We now know that the smallest eigenvalue, λ_σ, of the problem $L_u w = \lambda \sigma w$, satisfies $\lambda_\sigma \leq 2\sigma^{-1} \lambda_1$ and we let w_1 be a corresponding non-negative eigenfunction. Then $L_u^{-1} a(x) w_1 \geq L_u^{-1} \sigma w_1 = \lambda_\sigma^{-1} w_1$ and for any positive integral power ν, $(L_u^{-1} a(x))^\nu w_1 \geq \lambda_\sigma^{-\nu} w_1$. It follows that $\| (L_u^{-1} a(x))^\nu w_1 \|_0 \geq \lambda_\sigma^{-\nu} \| w_1 \|_0$ and hence

$$\| (L_u^{-1} a(x))^\nu \|^{1/\nu} \geq \lambda_\sigma^{-1}; \quad \nu = 1, 2, 3 \ldots \tag{2.7}$$

where the operator norm is that of a map from $\overset{\circ}{C}{}^0(\bar{\Omega})$ to itself. It now follows from (2.7) and the spectral radius formula ([14], p. 567) that

$$\lambda_a^{-1} = \lim_{\nu \to \infty} \| (L_u^{-1} a(x))^\nu \|^{1/\nu} \geq \lambda_\sigma^{-1}$$

and hence

$$\lambda_a \leq \lambda_\sigma \leq \frac{2\lambda_1}{\sigma}.$$

We can now prove a general existence theorem for sublinear elliptic problems. Related results have been obtained by H. Kuiper [9], [15], and by H. Amann (cf. [16] for references).

Theorem 2.5. Let L_u be the elliptic operator described in Lemma 2.4. Suppose $h = h(x, p, q)$ is continuously differentiable on $\bar{\Omega} \times R \times R^n$ and satisfies

$$\lim_{p \to 0} p^{-1} h(x, p, q) = +\infty \tag{2.8}$$

uniformly in x and q. Let S' be the set of solutions (λ, u) of

$$\left. \begin{array}{lll} L_u u = \lambda h(x, u, \nabla u) & & x \in \Omega \\ u > 0 & & x \in \Omega \end{array} \right\} \tag{2.9}$$

with $u \in C^2(\Omega) \cap \overset{\circ}{C}{}^0(\bar{\Omega})$ and let $S'' = (0, 0) \cup S'$ with the $R^+ \times C^{1+\alpha}(\bar{\Omega})$ topology. Then the component C of $(0, 0)$ in S'' is unbounded.

Proof. If $h(x, 0, 0)$ is not identically zero, then (2.9) may be rewritten so that Theorem 2.3 will apply. Since $p^{-1}h(x, p, q)$ is positive for p in some interval, $0 < p \leq p'$, we can write $h = h^+ - h^-$ with $h^{\pm} \geq 0$ and $h^-(x, p, q) = 0$ for $0 \leq p \leq p'$. Then the equation in (2.9) can be written

$$(L_u + \lambda \frac{h^-(x, u, \nabla u)}{u})u = \lambda h^+(x, u, \nabla u)$$

or as $u = \lambda F(\lambda, u)$ with

$$F(\lambda, u) = (L_u + \lambda \frac{h^-(x, u, \nabla u)}{u})^{-1} h^+(x, u, \nabla u).$$

Since $h^- = 0$ for $0 \leq p \leq p'$ one easily sees that for $\|u\|_{1+\alpha} \leq p'$ $F(\lambda, u) = L_u^{-1}h^+$ and that $F(\lambda, 0) = L_0^{-1}h^+(x, 0, 0) \neq 0$. Applying Theorem 2.3, we obtain the desired continuum C.

Now suppose $h(x, 0, 0) = 0$ for all $x \in \Omega$. Let η be the function introduced in the proof of theorem 2.1 and consider the problem

$$L_u u = \lambda[(1-\eta(n\|u\|))nu(x) + \eta(n\|u\|)h(x, u, \nabla u)] \tag{2.10}$$

for $u \in C^2(\Omega) \cap \mathring{C}^0(\bar{\Omega})$, where $\|u\| = \|u\|_{1+\alpha}$. The problem (2.10) has the form treated in Theorem 1.2 and from that theorem and the discussion following its proof we conclude that (2.10) possesses an unbounded continuum C_n of positive solutions emanating from $(\lambda_1^n, 0)$, where $\lambda_1^n = n^{-1}\lambda_1$, and λ_1 is the lowest eigenvalue of

$$L_0 w = \lambda w \tag{2.11}$$

If a point $(\lambda, 0)$ is in C, then λ must be an eigenvalue of $L_0 w = \lambda n w$ with $w > 0$ in Ω (cf. [1], p. 207). However, as we have observed, such a problem possesses a unique positive eigenvector, and we conclude that $\lambda = \lambda_1^n$. Since $h(x, 0, 0) = 0$ one verifies as in the proof of Theorem 2.1, writing (2.10) in the form (1.10), that $C = \lim\sup C_n$ consists of solutions of problem (2.9). Since $(0, 0) \in C$, to obtain the conclusion of the theorem we need only show that no point $(\bar{\lambda}, 0)$ with $\bar{\lambda} > 0$ can be in C. If there were such a point we could assume, after renumbering, that each C_n contained a pair (λ_n, u_n) and that (λ_n, u_n) converged to $(\bar{\lambda}, 0)$ as $n \to \infty$. Then the linear problem

$$L_{u_n} v = \lambda[(1-\eta(n\|u_n\|))n + \eta \cdot \frac{h(x, u_n(x), \nabla u_n(x))}{u_n(x)}]v \tag{2.12}$$

would have a solution $\lambda = \lambda_n$, $v = u_n$. We may assume that $\|u_n\|_0$ is small enough so that $h/u_n \geq 0$ and so that the estimate of Lemma 2.4 is applicable. As $u_n > 0$ in Ω, λ_n is the smallest eigenvalue of the problem (2.12). But for n sufficiently large, $n > 4\lambda_1/\bar{\lambda}$ and $h/u_n > 4\lambda_1/\bar{\lambda}$, implying that the function in square brackets in (2.12) is greater than $4\lambda_1/\bar{\lambda}$. Letting $\sigma = 4\lambda_1/\bar{\lambda}$ in Lemma 2.4, we see that $\lambda_n \leq \bar{\lambda}/2$ for all sufficiently large n, a contradiction.

3. Superlinear Operators

We begin this section with a brief discussion of an identity due to Pohozaev [17], from which one can see that there will be no "superlinear" theorems of the degree of generality of the theorems in section 2. As such we will be treating specific boundary value problems and will continue to restrict attention to classical Dirichlet problems.

Suppose $\Omega \subset R^n$ is as before, and let f be a real-valued function of a real variable. Let $F(t) = \int_0^t f(s)ds$. Suppose u satisfies

$$\left.\begin{aligned} -\Delta u &= f(u) \qquad x \in \Omega \\ u &= 0 \qquad x \in \partial\Omega \end{aligned}\right\} \tag{3.1}$$

and that $u > 0$ in Ω. Then

$$2n \int_\Omega F(u(x))dx + (2-n) \int_\Omega f(u(x))u(x)dx = \int_{\partial\Omega} \left|\frac{\partial u}{\partial \nu}\right|^2 r \cdot \nu d\sigma$$

where $\partial u/\partial\nu$ is the derivative of u in the direction of the outward normal at a point of $\partial\Omega$ and $r = (x_1, x_2, \ldots, x_n)$. If Ω is convex or, more generally, star-shaped, then with the vector r measured from a suitable origin, $r \cdot \nu \geq 0$ for every point of $\partial\Omega$. For such an Ω and for $f(s) = s^\beta$ $\beta > 0$, the identity (3.2) yields the inequality

$$\frac{2n}{\beta+} \int_\Omega u^{\beta+1}(x)dx + (2-n) \int_\Omega u^{\beta+1}(x)dx \geq 0$$

or

$$\beta \leq \frac{n+2}{n-2} \tag{3.3}$$

Thus, for $\beta > \frac{n+2}{n-2}$ there can be no positive solution of $-\Delta u = u^\beta$ with $u = 0$ on $\partial\Omega$. For $1 < \beta < \frac{n+2}{n-2}$, the existence of positive solutions, among others, follows from the work of Ambrosetti and Rabinowitz [18], who treat a wide class of elliptic equations having variational structure. Let $|r| = (\Sigma\, x_i^2)^{1/2}$ and suppose Ω is the annulus: $\{r = (x_1 \ldots x_n) \mid 0 < r_0 < |r| < r_1\}$. Then one can show the existence of many radial solutions of radially symmetric equations in Ω as a straightforward application of results from [19] and [20]. In particular, one has the following result in which there is no upper limit on the growth of h in the variable u.

<u>Theorem</u> 3.1. Suppose Ω is an annulus in R^n. Let $h = h(|r|, p, q)$ be continuous in $[r_0, r_1] \times R^+ \times R^n$ and satisfy:

i) $\lim_{p \to 0} p^{-1} h(|r|, p, q) = 0$, uniformly in $|r|$ and q

ii) $\lim_{p \to \infty} p^{-1} h(|r|, p, q) = +\infty$, uniformly in $|r|$ and q.

iii) $\lim_{|q| \to \pm\infty} |q|^{-2} h(|r|, p, q) = 0$, uniformly in $|r|$ and p.

Then for each $\epsilon > 0$, the problem

$$\left.\begin{aligned} -\Delta u &= \lambda h(|r|, u, \nabla u) && r \in \Omega \\ u &> 0 && r \in \Omega \\ u &= 0 && r \in \partial\Omega \end{aligned}\right\} \tag{3.4}$$

possesses positive solutions forming a continuum C in $R^+ \times C^1(\bar{\Omega})$ such that the projection of C on the λ axis is $[\epsilon, \epsilon^{-1}]$.

One can see from the discussion thus far that the geometry of Ω plays a role, not yet understood, in the question of existence of solutions.

We wish to establish the existence of positive solutions for a class of superlinear elliptic problems which are neither variational nor reducible to ordinary differential equations. Suppose that Ω is a simply-connected bounded domain in R^2 with a $C^{2+\alpha}$ boundary and consider the problem

$$\begin{aligned} -\Delta u &= \lambda g(x, u, \nabla u) & x \in \Omega \\ u &= 0 & x \in \partial\Omega \end{aligned} \qquad (3.5)$$

where, for $x \in \bar{\Omega}$, $p \geq 0$, and $q \in R^2$, $g(x, p, q)$ is continuous and satisfies

$$0 < A \leq p^{-\beta} g(x, p, q) \leq B < \infty \qquad (3.6)$$

for constants A, B, and β, with $\beta > 1$. An essential ingredient in obtaining positive solutions of (3. 5) is the existence of a priori bounds. We give a brief account of how one establishes bounds, referring the reader to [21] for details.

Since the domain Ω has a $C^{2+\alpha}$ boundary, a conformal map ϕ from Ω to the unit disc $D = \{\alpha = (\alpha_1, \alpha_2) \mid \alpha_1^2 + \alpha_2^2 < 1\}$ will have a Jacobian $d\phi/d\alpha$ satisfying

$$0 < \gamma \leq \left|\frac{d\phi}{d\alpha}\right| \leq \gamma^{-1} \qquad (3.7)$$

on the closed set $\bar{\Omega}$.

Using the conformal map to introduce a change of variables in (3. 5), one obtains an equation of the same form as (3. 5), in D, with a function

$\tilde{g}$ satisfying condition (3. 6) with constants $0 < \tilde{A} \leq \tilde{B} < \infty$. Hence, for the purpose of obtaining a priori bounds in C^0, we will assume, without loss of generality, that Ω in (3. 5) is D. We identify $x = (x_1, x_2)$ with $x_1 + ix_2$, α with $\alpha_1 + i\alpha_2$, and use $d\alpha = d\alpha_1 d\alpha_2$, $dx = dx_1 dx_2$. We use the Green's function $G = G(x, \alpha)$ for $-\Delta$ on D; namely,

$$G(x, \alpha) = -\frac{1}{2\pi} \log \left| \frac{x - \alpha}{1 - \bar{\alpha}x} \right| ,$$

where $|\ |$ is used for the modulus of a complex number. The a priori estimates are based on the following three lemmas.

<u>Lemma</u> 3. 2. Suppose $m \geq 1$, $s \geq 0$, and $2-m \leq sm < 3/2$. Then there is a constant K_1, depending upon s and m, such that

$$\int_D \frac{G^m(x, \alpha)}{(1-|\alpha|)^{sm}} d\alpha \leq K_1 (1-|x|)^{2-sm} . \tag{3. 8}$$

<u>Lemma</u> 3. 3. Let f be a continuous function defined on $D \times [0, \infty)$, satisfying

$$Au^\beta \leq f(x, u) \leq B \max(C^\beta, u^\beta)$$

for constants $A > 0$, $B > 0$, $C \geq 0$, and $\beta > 1$. There is a constant $K_2 = K_2(A, B, C, \beta)$ such that if $u \geq 0$ satisfies

$$\left.\begin{aligned} -\Delta u &= f(x, u); \quad x \in D \\ u &= 0 \qquad\qquad x \in \partial D \end{aligned}\right\} \tag{3. 9}$$

and $u(x_0) = \max_{x \in D} u(x) = M$, with $M \geq C$, then

$$1 - |x_0| \leq K_2 M^{(1-\beta)/2} \tag{3. 10}$$

Lemma 3.4. Let f be as in the previous lemma, and suppose $u \geq 0$ is a solution of (3.9). Let $J_0 = J_0(x)$, with $\int_D J_0 dx = 1$, be an eigenfunction corresponding to the lowest eigenvalue of $-\Delta J = \lambda J$ on D with $J = 0$ on ∂D. Then

$$\int_D J_0(x) u^\beta(x) dx \leq K_3 \tag{3.11}$$

where K_3 depends only on A, B, C, and β.

Theorem 3.5. Let Ω be a bounded, simply-connected domain in R^2 with $C^{2+\alpha}$ boundary $\partial\Omega$. Suppose $f = f(x, u)$ is continuous on $\Omega \times R^+$ and satisfies

$$Au^\beta \leq f(x, u) \leq B \max(C^\beta, u^\beta)$$

for constants $A > 0$, $B > 0$, $C \geq 0$ and β satisfying $1 < \beta < 3$. Then there is a constant K depending only on Ω, A, B, C, and β such that, if $u \in C^2(\Omega) \cap \mathring{C}^0(\bar{\Omega})$, $u \geq 0$, and

$$-\Delta u = f(x, u), \quad x \in \Omega \tag{3.12}$$

then

$$\|u\|_0 \leq K . \tag{3.13}$$

Proof. We treat the case in qhich $\Omega = D$ and $f(x, u) = u^\beta$. The essentials of the proof can be seen in that case. A solution of (3.12) will satisfy

$$u(x) = \int_D G(x, \alpha) u^\beta(\alpha) d\alpha .$$

Suppose that $u(x_0) = \|u\|_0 = M$ and let $1 - |x_0| = \delta$. Then

$$
\begin{aligned}
u(x_0) &\leq M^{\beta/2} \int_D \frac{G(x_0, a)}{J_0^{1/2}(a)} J_0^{1/2}(a) u^{\beta/2}(a) da \\
&\leq M^{\beta/2} \left(\int_D \frac{G^2(x_0, a)}{J_0(a)} da \right)^{1/2} \left(\int_D J_0(a) u^{\beta}(a) da \right)^{1/2}
\end{aligned}
\tag{3.14}
$$

The fact that $J_0(a) \geq 1 - |a|$ is easily established from known results on Bessel functions, so we may use lemma 3.2 with $m = 2$ and $s = 1/2$, together with lemma 3.4, to conclude from (3.14) that

$$u(x_0) \leq M^{\beta/2} K_1^{1/2} \delta^{1/2} K_3^{1/2} .$$

Then, from lemma 3.3, we see that $\delta \leq K_2 M^{(1-\beta)/2}$. Since $u(x_0) = M$,

$$
\begin{aligned}
M &\leq M^{\beta/2} K_1^{1/2} K_2^{1/2} M^{\frac{1-\beta}{4}} K_3^{1/2} \\
&= M^{\frac{\beta+1}{4}} K_4
\end{aligned}
$$

yielding

$$M \leq K$$

where $K = K_4^{4/(3-\beta)} = (K_1 K_2 K_3)^{2/(3-\beta)}$.

If one keeps track of constants in the general problem, noting (3.7), one can show that

$$
K = \left[(1+2^6) 2^{\frac{1}{4}} \gamma^{\frac{3}{2}} B^{\frac{3}{4}} e^{\gamma^2 B^{-\frac{1}{2}} A^{\frac{1}{2}}} \left(2 \left(\frac{12\gamma^2}{A} \right)^{\frac{\beta}{\beta-1}} + C^{\beta} + \frac{5\gamma^2 C}{A} \right)^{\frac{1}{2}} + \frac{\gamma^2 BC}{8}^{\frac{3\beta-1}{4}} \right]^{\frac{4}{3-\beta}} + 4C \tag{3.15}
$$

werves as a bound in the theorem.

Note that if one has a radial solution $u = u(|x|) \geq 0$ of a problem such as (3.12) in $\Omega = D$, then one obtains an a priori bound for any value of $\beta > 1$. This follows from lemma 3.3 since, in the radial case, the maximum of u occurs at $x = 0$ (cf. [21]).

R. Nussbaum in [22] has independently obtained bounds for elliptic problems, for the purpose of obtaining existence results. He treats the Dirichlet and other boundary value problems for a second order operator in divergence form. In the situation of Theorem 3.5 his results give bounds when $1 < \beta < 2$.

Using the explicit form for K in (3.15), one obtains the following estimate.

<u>Lemma</u> 3.6. Let $\Omega \subset R^2$ be as above and suppose $g = g(x, p, q)$ satisfies (3.6) with $1 < \beta < 3$. Then there is a constant $\hat{K} = \hat{K}(\Omega, A, B, \beta)$ such that for any positive integer n, a solution $u \geq 0$ of

$$\left.\begin{aligned} -\Delta u &= \lambda(n^{-1}u + g(x, u, \nabla u)); && x \in \Omega \\ u &= 0 && x \in \partial\Omega \end{aligned}\right\} \tag{3.16}$$

must satisfy

$$\|u\|_0 \leq \frac{\hat{K}}{\lambda^{1/(\beta-1)}} \tag{3.17}$$

<u>Proof.</u> Let λ_0 be the lowest eigenvalue of $-\Delta w = \lambda w$. Since the linear problem

$$-\Delta w = \lambda(n^{-1} + \frac{g(x, u(x), \nabla u(x))}{u(x)})w \tag{3.18}$$

is satisfied by $u = w$ and since $g/u \geq 0$, it follows from a variational characterization of the lowest eigenvalue of (3.18) that it will be no greater than the lowest eigenvalue of $-\Delta w = \lambda n^{-1} w$, i.e., $\lambda \leq n\lambda_0$. If we let

$$f(x, p) = \lambda(n^{-1}p + g(x, p, \nabla u(x))) \tag{3.19}$$

then

$$\lambda A u^{\beta} \leq f(x, u) \leq \lambda_0 u + \lambda B u^{\beta}. \tag{3.20}$$

The terms $\lambda_0 u$ and $\lambda B u^{\beta}$ are equal for $u = u_{\lambda} = (\lambda_0 \lambda^{-1} B^{-1})^{\frac{1}{\beta-1}}$ and one readily checks that

$$\lambda_0 u + \lambda B u^{\beta} \leq 2\lambda B \max(u_{\lambda}^{\beta}, u^{\beta}). \tag{3.21}$$

Thus $f(x, p)$ defined by (3.19) satisfies the hypotheses of heorem 3.5 with A replaced by λA, with B replaced by $2\lambda B$, and with $C = u_{\lambda}$. Using these new constants in (3.15), one finds that $\lambda^{-1/(\beta-1)}$ factors out of K, yielding the desired bound.

We can now obtain an existence theorem for superlinear equations. For the statement we add "∞" to R^{+} and to the Banach space $C^{1+\alpha}(\bar{\Omega})$, as earlier, by the standard device of the stereographic projection with the corresponding induced metric.

<u>Theorem</u> 3.7. Let Ω and $g = g(x, p, q)$ be as in lemma 3.7 and suppose further that g is a C^{1} function. Let S' denote the set of solutions (λ, u) of

$$\left.\begin{aligned} -\Delta u &= \lambda g(x, u, \nabla u) & x \in \Omega \\ u &= 0 & x \in \partial\Omega \end{aligned}\right\} \tag{3.22}$$

with $u > 0$ in Ω. Then, with the $(R^{+} + \infty) \times (C^{1+\alpha}(\bar{\Omega}) + \infty)$ topology on S', the set $S' \cup (0, \infty) \cup (\infty, 0)$ contains a continuum C which includes $(0, \infty)$ and $(\infty, 0)$. Moreover, for $0 < \lambda < \infty$, $(\lambda, u) \in C$ satisfies

$$\|u\|_{2+\alpha} \leq \tilde{K}\lambda^{-\frac{1}{\beta-1}} \tag{3.23}$$

for some constant $\tilde{K} = \tilde{K}(\Omega, A, B, \beta)$.

Proof. For each positive integer n, the problem (3.16) satisfies the conditions of Theorem 1.2 with $m = 1$ and $\lambda_1 = n\lambda_0$ where λ_0 is the first eigenvalue of $-\Delta$ on Ω. We conclude that problem (3.16) possesses an unbounded continuum C_n of positive solutions emanating from $(n\lambda_0, 0)$. As in the proof of theorem 2.5, one sees that $(n\lambda_0, 0)$ is the only trivial solution of (3.16) in C_n.

Let K denote the cone of nonnegative functions in $\overset{\circ}{C}{}^{1+\alpha}(\bar{\Omega})$. Using the fact that $(-\Delta)^{-1}$ maps bounded sets in $C^{\alpha}(\bar{\Omega})$ to precompact sets in $C^{1+\alpha}(\bar{\Omega})$, one shows that the collection of all nonnegative solutions of (3.22), and of (3.16) for all values of n, constitute a precompact subset of $(R^+ + \infty) \times (K + \infty)$. In the proof of lemma 3.6, we have seen that $(\lambda, u) \in C_n$ implies $\lambda \leq n\lambda_0$. If we establish (3.23) for $(\lambda, u) \in C_n$, then since C_n is unbounded it must contain $(0, \infty)$. Let $\tilde{C}_n = C_n \cup \{[n\lambda_0, \infty] \times \{0\}\}$. According to [3], lemma 1.9, the set $C = \limsup \tilde{C}_n$ is a continuum in $(R^+ + \infty) \times (K + \infty)$. Clearly $(0, \infty)$ and $(\infty, 0)$ are in C and, assuming (3.23) for C_n, it follows for points in C with $0 < \lambda < \infty$. Since $(-\Delta)^{-1}(n^{-1}u + g)$ converges in $C^{1+\alpha}(\bar{\Omega})$ to $(-\Delta)^{-1}g$, uniformly on bounded sets in K, all points in C with $0 < \lambda < \infty$ are solutions of (3.22). Provided there are no trivial solutions in C, C will satisfy the conclusions of the theorem. Hence it remains to be shown that (3.23) holds for positive solutions of (3.16) and that C contains no point

$(\lambda, 0)$ with $0 \leq \lambda < \infty$.

To obtain (3.23) from the bound (3.17) one can appeal to the L^p theory of elliptic equations (cf. [11]). Omitting the factor $\lambda^{-1/(\beta-1)}$, one sees that $\|u_0\|_0 \leq \hat{K}$ implies $\|n^{-1}u + g\|_{L^p} \leq \hat{K}'$, using (3.6). Then in the Sobolev space $\overset{\circ}{W}{}^2_p(\Omega)$, one has u, the solution of (3.16), satisfying $\|u\|_{2,p} \leq \hat{K}''$. For p large, one then obtains $\|u\|_{1+\alpha} \leq \hat{K}'''$ from embedding theorems (cf. [12], Lecture II) and as one now has a bound on $n^{-1}u + g(x, u, \nabla u)$ in C^α, a bound $\|u\|_{2+\alpha} \leq \tilde{K}$ follows from the Schauder estimates (cf. [11]).

As regards C containing trivial solutions, suppose there were a sequence $(\lambda_{n_k}, u_{n_k}) \in C_{n_k}$, converging to $(\lambda, 0)$ in $R^+ \times C^{1+\alpha}(\bar{\Omega})$, as $k \to \infty$. We could renumber the sequence and obtain

$$-\Delta \frac{u_n}{\|u_n\|_0} = \lambda_n\left(\frac{u_n}{n\|u_n\|_0} + g\frac{(x, u_n, \nabla u_n)}{\|u_n\|_0}\right) \tag{3.24}$$

from equation (3.16), with $\lambda_n \to \lambda$ and $\|u_n\|_0 \to 0$, as $n \to \infty$. The right hand side of (3.24) converges to 0 in $C^0(\bar{\Omega})$ as $n \to \infty$ and hence it converges to zero in $L^p(\Omega)$ for any $p \geq 1$. As in the last paragraph, one sees that for p sufficiently large, the solution $u_n/\|u_n\|_0$ will converge to zero in $C^0(\bar{\Omega})$. As the solution has norm 1 for all n, we have a contradiction.

REFERENCES

1. Krasnoselskii, M. A., Positive Solutions of Operator Equations, Noorhoff, Groningen, 1964.

2. Krein, M. G. and Rutman, M. A., Linear operators leaving invariant a cone in a Banach space, Amer. Math. Soc. Transl. Ser. 1, Volume 10, 1962.

3. Turner, R. E. L., Transversality and cone maps, to appear in Arch. Rat. Mech. Anal.

4. Turner. R. E. L., Transversality in nonlinear eigenvalue problems, Contributions to Nonlinear Functional Analysis, edit. E. Zarantonello, pp. 37-68, Academic Press, New York, 1971.

5. Dancer. E. N., Global solution branches for positive mappings, Arch. Rat. Mech. Anal. 52(1973), 181-192.

6. Ladyzhenskaya, O. A. and Uraltseva, N., Linear and Quasilinear Elliptic Equations, Academic Press, New York, 1968.

7. Rabinowitz, P. H., A global theorem for nonlinear eigenvalue problems and applications, Contributions to Nonlinear Functional Analysis, Edit. E. Zarantonello, pp. 11-36, Academic Press, New York, 1971.

8. Whyburn, G. T., Analytic Topology, Amer. Math. Soc. Colloq. Publ., Vol. 28, Providence, 1942.

9. Kuiper, H. J., On positive solutions of nonlinear elliptic eigenvalue problems, Rend. Cir. Mat. di Palermo, Ser. II, Vol. 20 (1971), 113-138.

10. Stampacchia, G., Le problème de Dirichlet pour les équations elliptiques du second ordre à coefficients discontinus, Ann. Inst. Fourier 15 (1965), 189-258.

11. Agmon, S., Douglis, A., and Nirenberg, L., Estimates near the boundary for solutions of elliptic partial differential equations satisfying general boundary conditions, I, Comm. Pure Appl. Math. 12(1959), 623-727.

12. Nirenberg, L., On elliptic partial differential equations, Ann. Scuola Norm. Sup. Pisa, Ser. 3, 13 (1959), 1-48.

13. Kato, T., Perturbation Theory for Linear Operators, Springer-Verlag, New York, 1966.

14. Dunford, N. and Schwartz, J., Linear Operators, Vol. 1, Interscience New York, 1958.

15. Kuiper, H. J., Eigenvalue problems for noncontinuous operators associated with quasilinear elliptic equations, Arch. Rat. Mech. Anal. 53(1974), 178-186.

16. Amann, H., Multiple positive fixed points of asymptotically linear maps, cf. these proceedings.

17. Pohozaev, S-I., Eigenfunctions of the equation $\Delta u + \lambda f(u) = 0$, Soviet Math., Vol. 6 (1965), 1408-1411.

18. Ambrosetti, A. and Rabinowitz, P., Dual variational methods in critical point theory and applications, J. Funct. Anal. 14(1973), 349-381.

19. Turner, R. E. L., Superlinear Sturm-Liouville problems, J. Diff. Equations 13(1973), 157-171.

20. Kuiper, H. J. and Turner, R. E. L., Sturm-Liouville problems with prescribed nonlinearities, to appear.

21. Turner, R. E. L., A priori bounds for positive solutions of nonlinear elliptic equations in two variables, Duke Math. J., to appear.

22. Nussbaum, R., Positive solutions of some nonlinear elliptic boundary value problems, to appear.